Elements of STIL

Principles and Applications of IEEE Std. 1450

FRONTIERS IN ELECTRONIC TESTING

Consulting Editor
Vishwani D. Agrawal

Books in the series:

Elements of STIL: Principles and Applications of IEEE Std. 1450
G. Maston, T. Taylor, J. Villar
ISBN: 1-4020-7637-1

Fault Injection Techniques and Tools for Embedded systems Reliability Evaluation
A. Benso, P. Prinetto
ISBN: 1-4020-7589-8

High Performance Memory Memory Testing
R. Dean Adams
ISBN: 1-4020-7255-4

SOC (System-on-a-Chip) Testing for Plug and Play Test Automation
K. Chakrabarty
ISBN: 1-4020-7205-8

Test Resource Partitioning for System-on-a-Chip
K. Chakrabarty, Iyengar & Chandra
ISBN: 1-4020-7119-1

A Designers' Guide to Built-in Self-Test
C. Stroud
ISBN: 1-4020-7050-0

Boundary-Scan Interconnect Diagnosis
J. de Sousa, P.Cheung
ISBN: 0-7923-7314-6

Essentials of Electronic Testing for Digital, Memory, and Mixed Signal VLSI Circuits
M.L. Bushnell, V.D. Agrawal
ISBN: 0-7923-7991-8

Analog and Mixed-Signal Boundary-Scan: A Guide to the IEEE 1149.4 Test Standard
A. Osseiran
ISBN: 0-7923-8686-8

Design for At-Speed Test, Diagnosis and Measurement
B. Nadeau-Dosti
ISBN: 0-79-8669-8

Delay Fault Testing for VLSI Circuits
A. Krstic, K-T. Cheng
ISBN: 0-7923-8295-1

Research Perspectives and Case Studies in System Test and Diagnosis
J.W. Sheppard, W.R. Simpson
ISBN: 0-7923-8263-3

Formal Equivalence Checking and Design Debugging
S.-Y. Huang, K.-T. Cheng
ISBN: 0-7923-8184-X

Defect Oriented Testing for CMOS Analog and Digital Circuits
M. Sachdev
ISBN: 0-7923-8083-5

Reasoning in Boolean Networks: Logic Synthesis and Verification Using Testing Techniques
W. Kunz, D. Stoffel
ISBN: 0-7923-9921-8

Introduction to I_{DDQ} Testing
S. Chakravarty, P.J. Thadikaran
ISBN: 0-7923-9945-5

Elements of STIL

Principles and Applications of IEEE Std. 1450

by

Gregory A. Maston

Tony R. Taylor

Julie N. Villar

Synopsys, Inc.

Kluwer Academic Publishers
Boston/Dordrecht/London

Distributors for North, Central and South America:
Kluwer Academic Publishers
101 Philip Drive
Assinippi Park
Norwell, Massachusetts 02061 USA
Telephone (781) 871-6600
Fax (781) 871-6528
E-Mail <kluwer@wkap.com>

Distributors for all other countries:
Kluwer Academic Publishers Group
Post Office Box 322
3300 AH Dordrecht, THE NETHERLANDS
Telephone 31 78 6576 000
Fax 31 78 6576 474
E-Mail <orderdept@wkap.nl>

Electronic Services <http://www.wkap.nl>

Library of Congress Cataloging-in-Publication

CIP info or:

Title: Elements of STIL
Principles and Applications of IEEE Std. 1450
Author (s): Gregory A. Maston, Tony R. Taylor and Julie N. Villar
ISBN: 1-4020-7637-1

Printed on acid-free paper.

Printed in the United States of America

Contents

Standard
Test Interface
Language

List of Figures

Standard
Test Interface
Language

List of Tables

Standard Test Interface Language

Preface

Standard
Test Interface
Language

So I was wrong. I was absolutely sure that by having an IEEE Standard defined, reviewed, and accepted, that I wouldn't need to write a book about it as well. The Standard would be the complete reference. And be aware — this book does not serve as a replacement to the IEEE Std. 1450 document. You should have a copy of the Standard as you go through this book.

I realized that the Standard would not be the complete reference, about the time that the Working Group started to put notes into the draft proposal — notes to elaborate decisions in the Working Group, but that would be removed in the final draft. Then, once the Standard was accepted I became the central point of contact for people who just picked up the Standard, who didn't have the benefit of the Working Group discussions, who only had available that one final sentence in the Standard and who didn't benefit from the perspective of where those words came from. Sometimes those questions have resulted in clarifications to the Standard. Sometimes I would respond to those questions with more background and perspective as well.

It is this additional background and perspective I hope you find in this book. Not the "what" of STIL, but the "why". Included are the motivations of the Working Group, with the hope that understanding these motivations provides an appreciation of the proper application and an awareness of the improper application of these constructs. And a perspective on how to apply STIL for you.

I started in the STIL effort as both a user and a provider of another test interface language, one of the interfaces that was congesting the industry with a slurry of local formats. My hope with the STIL effort was to create an official standard that would not support a larger set of requirements. Currently I'm still a provider, but now purely on the EDA side and responsive to those users of whom I use to be one.

Greg Maston

I remember being asked at ITC one year whether STIL would be a standard in our life time. This was, and still is, a shared concern by many of us who continue to work through all the issues.

The thing that has kept me going for almost 10 years is the absolute conviction that this standard is as necessary to our industry today as it was when 100 concerned ATE users began this initiative in 1994.

As with all standards, coming to consensus requires compromise. This standard does not completely meet the expectations of each user for every application. But, with the latest extensions, it is becoming a very rich language that is enabling new applications that are taking design and test to new levels.

Tony Taylor

The Standard Test Interface language (STIL) development started in 1994 by a group of people interested in transporting test data from CAD generation environments to ATE environments. The effort has been evolving ever since. This language represents a new direction in the evolution of a language to support testers. The recognized need for managing the large volumes of (ATPG) test data for digital designs has driven the development of STIL.

Moreover the acceptance process for a new language is slow, especially for a language as affluent in features as STIL. The adoption process is dependent on the dissemination of the language and how to use it. To date, the Standard is the only available information to the design community to explain what

STIL is and the uses of it's constructs from the EDA environment to the Tester environment. This is the only book that elaborates on IEEE Std. 1450-1999.

The development and architecture of Synopsys' product TetraMAX ATPG® has motivated and accelerated the need for STIL acceptance with partners, competitors, ATE corporations, 3rd party interfaces and other interoperability products at Synopsys. It is one of the most compelling reasons today leading the adoption of STIL.

Both Synopsys and the authors are committed to the widespread use of STIL as a software [EDA] to hardware [ATE] description language. This book should be read by those designing for test, supporting the testing of Integrated Circuits, transporting test data to and from CAD generation environments onto test environments, and supporting digital test applications.

Julie Villar

Acknowledgements

Standards development relies on active participants, and I thank all of the members of the STIL Working Group, and the oversight committee, for their involvement. But also important is the corporate support to facilitate the process, and for that I thank Dan Perrine, and Motorola, and TSSI, and Synopsys, and all my co-workers for their valued inputs. And Elizabeth, and Nate and Nick, for their loving support during all these efforts.

Greg

I thank all my peers who have helped make me successful in the applications of STIL and supporting the interpretation of STIL. Special thanks to Tim Hunkler, Peter Wohl, Greg Maston, John Cosley, and the customers I support day to day over the last six years with STIL needs.
I thank my family for listening to the never ending soon, too busy, too tired, or too distracted excuses.
I thank Greg Maston for the work that he contributed to this book and his outstanding attitude he maintained while managing a full time career and this book in under a year. I've always been impressed with his amount of knowledge and now I am amazed at the end — you really made this book what it is.

Julie

CHAPTER 1

Foundations of STIL

Standard Test Interface Language

Understanding the history of the STIL effort provides an understanding of the language itself.

1.1 Introduction

The digital test domain has had a variety of standards to apply, but none that were defined explicitly with a goal of transportability to the test environment from Computer-Aided Design (CAD) contexts. The Standard Test Interface Language, identified by the acronym STIL and consciously mispronounced ***style*** by the Working Group, was defined with this purpose as a primary goal. However, as in many situations, misapplication of this standard can result in something that may not be directly applicable to a test environment.

This book presents an in-depth discussion of the STIL language and examples that demonstrate practical applications of the language. These presentations are reinforced with perspectives of the Working Group intentions. And finally, and most importantly, this book covers applications of STIL — the application of the language as it is intended to be used, and methodologies to redress the wrongs.

The complete definition of STIL is found in a document printed by the IEEE Standards Association and identified by the label *IEEE Standard 1450-1999*. This is citation [1] in the Bibliography. This book does not replace that Standard document, and it is recommended that you have a copy of the Standard available to review while using this book. When this book refers to "the Standard", this is the reference that is being made.

1.2 Organization of this Book

This Chapter covers the historical aspect of STIL — the underlying foundation resulting in a Standards effort, and provides an orientation on the motivations that resulted in the Standard. Chapter 2 extends this foundation with a discussion of the test environment that STIL is designed to address. The following chapters present the STIL constructs, their motivation as part of the language, combined with STIL applications, contexts of appropriate uses, and examples of inappropriate uses of STIL.

Terminology used in this book can be found in the Terminology section toward the end; scanning through this section quickly now might reduce subsequent confusion.

Special considerations for applying STIL concepts are highlighted under sections titled "STIL Conventions". These sections identify recommendations that are intended to maximize the utility of STIL as it was intended to be applied. These situations are not (generally) requirements of the language itself, but principles for using the language in the fashion it was intended.

Each convention has an icon in front of it. These icons have the following meanings:

 **To Tester**: this convention has an impact or consideration when the STIL data is being ported to be executed on ATE or in an ATE environment.

Interchange: this convention has a potential impact as STIL data is ported around to different environments.

Not STIL: this convention restates a construct that is not supported under STIL or elaborates on a restriction with a construct, with additional information about this restriction.

Additionally, the following icons appear in this book:

Background: the next sections cover some basic STIL syntax issues, test perspectives, or a concept not specifically related to the chapter topic but a good spot to present the issue. This icon indicates the next information provides perspective not immediately associated with the STIL constructs discussed in this chapter.

Close-up: the next sections cover a detail of the language for advanced understanding of the language.

1.3 History

The STIL effort was initiated in December 1994 at a Motorola hosted meeting in Austin, Texas. The first meeting was sponsored by the PowerPC division at Motorola, who organized a consortium of ATE vendors, EDA tool providers, and users of these tools, to participate in the definition of an interface language. It was intended that this interface language would be generated by the EDA tool providers and consumed by the ATE vendors directly — without intervening transforms. This organization of participants, shown in Figure 1.1, became known as the STIL triumvirate. As we shall see, participation by individuals in all three areas was central to the success of this definition.

This effort was initiated to respond to an issue that the Power PC program was experiencing. The Power PC effort was supported by divisions in both IBM and Motorola, and there was a problem between the two organizations. The two organizations were experiencing a bottleneck in the flow of test information between them. Each group had their own test equipment, and a common format at the test level was not available. The closest thing to a common format they could both support was an intermediate format that still needed to be transformed at each destination into specific test formats. The process of generating this intermediate format at the source, transmitting this format, reading this format at the destination, generating a program for the tester, and loading this program onto a tester was taking the better half of a day. The size of the test data was making a slow process even slower. But this wasn't the worse part of the problem: identifying problems in the test program, fixing those problems (in the intermediate format), regenerating the tester object, reloading the tester and testing the fix was also taking much too much time.

Motorola was looking for an immediate fix to this problem — a format that could be transferred between divisions, that could be loaded and executed directly on their respective test equipment directly, and that could be debugged in that format. The business issue for the success of the PowerPC program was crucial, and the strategy of leveraging the support of the triumvirate — particularly the users who would be primary in motivating both the CAD tool providers and the ATE vendors — was critical for this effort to be successful.

Standards development efforts do not move fast. First, there is the issue of defining the scope of the effort. Scope definition is a balancing act of defining just enough to support the requirements without defining too much, which slows down the Working Group efforts to complete the task at hand. Partition-

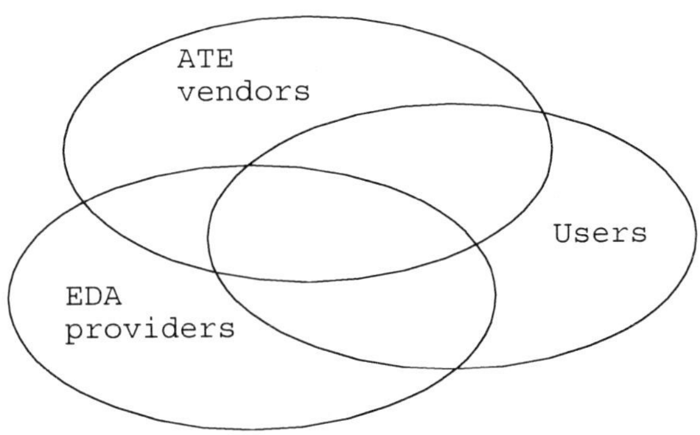

FIGURE 1.1 The Founding STIL triumvirate — a ruling organization of 3 parties

ing the scope is one way to allow the Working Group to focus on key aspects of the problem, and generate a standard with sufficient capability to address those problems. Partitioning is a negotiating process as well, which may result in follow-on projects to be defined (or at least promised); for the STIL effort this has resulted in an array of follow-on projects.

But even with the scope defined, the efforts of the Working Group, which make progress on a consensus basis, still take time. Generating an agreement among all members of a Working Group takes time; generating agreement amount the reviewers who do not have the benefit of the Working Group discussions takes more time. And to top it all off, all the members of the Working Group are volunteers who are investing their time as available. Figure 1.2 presents the timeline of the initial STIL effort.

The PowerPC program found solutions to address their bottleneck. But the motivation and interest in what they started continued on.

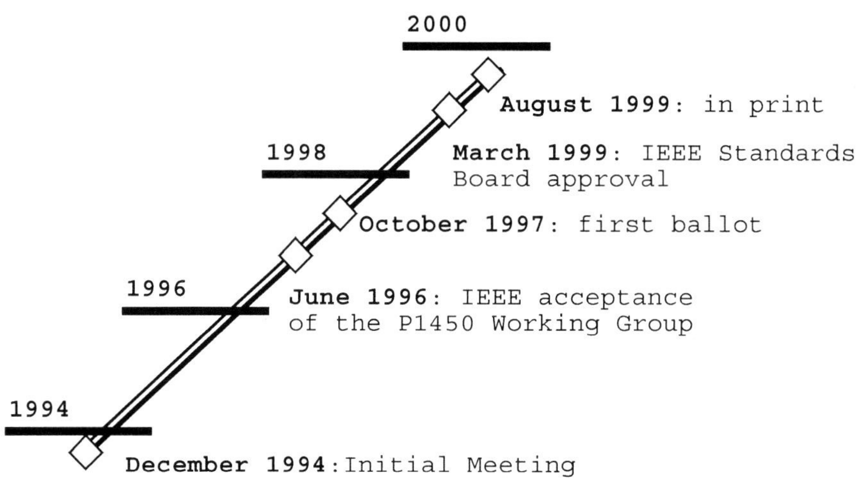

FIGURE 1.2 STIL Standard Development Timeline

1.4 Foundation Efforts

Several design flow and integrated circuit development environments contributed test requirements to the STIL development effort. Any individual contributor cannot be singled out; it speaks to the standards development process that all participants contribute to a consensus definition that is, ultimately, owned by all who contributed to the definition.

But the founding triumvirate provided some specific contributions to the effort. Several ATE organizations contributed languages and constructs to the definition. And on the Users side, several proposals were made of completely defined languages, particularly from the Application Specific Integrated Circuit, or ASIC, environments. ASIC Test, as discussed in the next chapter, had previously addressed some of the scope issues motivating this standard, in particular the requirement to transport a test program to multiple ATE platforms. All of these contributors are identified more fully in section 1.6. However, none of these proposals as they came to the Working Group satisfied the requirements that were being organized across all members of the effort. The effort to define a standard — and to reach consensus on a single definition — was still to come.

1.5 Scope of STIL and Follow-On Efforts

When defining the scope of a Standards effort, every application context or environmental connotation represents a requirement that may become part of the scope of that effort. Consider the group of people that came together to define this effort. The Triumvirate was founded by organizations with an interest in transporting test data from EDA generation environments onto test environments. This statement, in itself, represents a limitation on the effort and therefore an aspect to the scope definition. Full custom test development that does not rely on the bulk of test data coming from EDA sources had limited direct interest in this effort under this context.

Furthermore the bulk of the test data to transport from EDA to ATE consists generally of digital test data, and is typically generated to support testing of

ICs — at least IC manufacturers were a large part of the users in the founding Triumvirate, and IC-centric ATE manufacturers were also participating (at the impetus of their customers). Hence the scope of this effort was defined to support digital test application, for Integrated Circuits and their Assemblies.

This much of the scope was easily defined, at least by the participants in the effort at this initial phase. Certain words were carefully picked to be in the scope definition as well, in particular the term "bulk of the test data", to allow the Working Group to leverage identifying aspects of the test environment that could be considered "bulk" and allow the group to focus on the key aspects of the transportation issue and minimize distractions defining constructs that did not immediately affect the transportation issue.

1.5.1 Negotiating the Scope Limits

At this point, negotiations on the scope started. It is just as important to define what is outside the scope of the standard, as what is included. This is necessary to keep the focus on the key required features and to minimize efforts down stray paths. We found, just as many standards before us, that a standard for a limited set of features is more likely to succeed than a standard that attempts to cover too much ground. In fact, without a well defined focus, a document may never get through the balloting process that is critical for both validation of the effort as well as acceptance.

Features or behaviors that were negotiated out of this first development effort ended up in a list in Clause 1.1 of the Standard. Some of the negotiated points were still being discussed as the draft document entered balloting. Several of these issues immediately became additional efforts after the first Standard completed balloting. The timeline of Figure 1.3 identifies these efforts and they are individually outlined in Table 1.1. Some of these efforts have direct application to the existing Standard and are covered in this book as well.

1.5.2 Format or Language?

A more political discussion occurred early in the meetings, as part of the effort to establish the magnitude of the effort. This issue was encapsulated in a single, very long, discussion about whether this gathering would be defining a *format* or a *language*. The connotations of each of these terms was interpreted

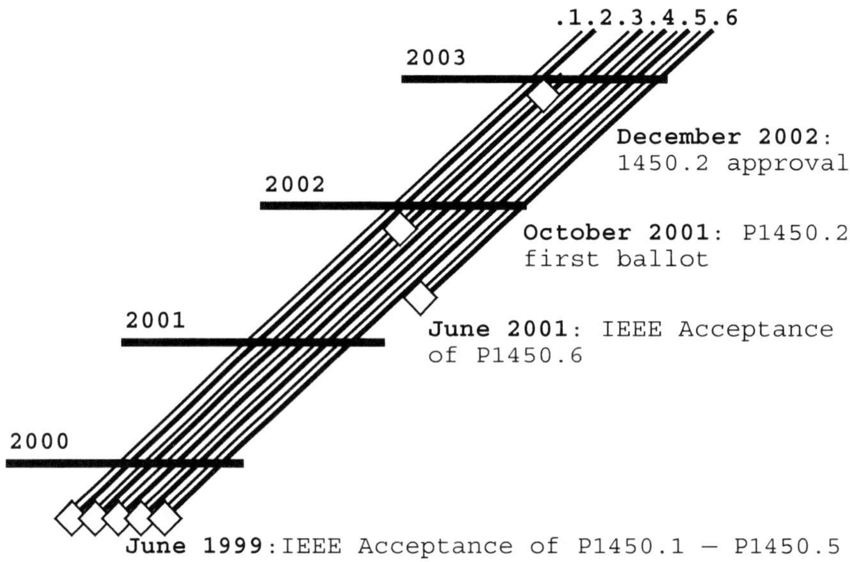

FIGURE 1.3 The Subprojects Timelines

to have impact on the overall effort. A *format* was seen to be of smaller overall effort, requiring less definition and completing in a shorter period of time. A *language* was seen to entail more rigorous definition and require correspondingly more development time.

This issue took a fair amount of discussion to resolve, and was ultimately resolved from two directions. The first was an identification that in order to define an adequate standard, the description process was equivalent whether defining a format or a language. The second direction had more of a marketing orientation: of the various acronyms proposed to name this effort, the Working Group didn't like naming the effort *STIF*.

1.5.3 Compression and the Binary Format Discussion

One rather integral discussion in the definition of the Standard involved the notion of defining a binary format of the test data. At least one of the formats proposed to the Working Group was a binary format, and the benefits of a binary format were well known to the Working Group, particularly as they directed affected the issue of "bulk of data". Several Working Group meetings featured a sub-group meeting to discuss the binary issue. This was one of those issues that the Working Group worked through difficult consensus, and which ended up with a rather detailed Annex D in the Standard to explain why the Standard ended up referencing an open-source compression package (gzip) to address the size issue — but not defining an integral binary format as part of the Standard.

1.5.4 Evolving Standards

Both the test environment and the design environment are constantly evolving. Circuit constructs, requiring new test mechanisms, are being constantly defined; test methodologies involving design constructs continue to be created as well. A line can be drawn around a subset of the existing contexts, and if that line is sufficiently encompassing, a Standard supporting those specific issues will be usable. That was very much the initial goal of this Standards effort. But the need to support additional constructs and capabilities was an integral part of the working definition as well.

As additional areas of the test problem were identified, follow-on efforts were defined. These efforts were initially defined to satisfy the concerns about scope limitations in the current work. The currently defined subprojects are identified in Table 1.1; some of these efforts are established standards at this time, others are still in the draft and Working Group review stage. Efforts labeled with a *P* are in Working Groups; efforts postfixed with a year identify the year that the effort became an official standard.

Figure 1.3 identifies the project inceptions and progress of these additional efforts.

TABLE 1.1 STIL Subprojects

IEEE ident.	Title/Purpose
P1450.1	*Extensions to STIL for Semiconductor Design Environments.* Supports EDA-specific application of STIL data, test development and test refinement processes. Defines constructs (notably Variables) applied in other subproject efforts as well, and as such is an important extension for additional STIL work.
1450.2-2002	*Extensions to STIL for DC Level Specification.* Identifies voltage, current, and other parametric relationships to apply or measure during test operation.
P1450.3	*Extensions to STIL for Tester Target Specification.* Provides mechanisms to identify limitations/constraints on STIL constructs as applied to specific test environments. Also used as a metric for an existing STIL test, to identify how much or how many elements are present.
P1450.4	*Extensions to STIL for Test Flow Specification.* Addresses test sequencing, organization, failure processing, and overall test control.
P1450.5	*Extensions to STIL for Semiconductor Test Method Specification.* An adjunct to the flow definition, provides definition for the constructs necessary to perform specific test operations.
P1450.6	*Extensions to STIL for Core Test Language (CTL) Support.* Defines constructs and test organizations to facilitate the reuse of test information generated for previously defined logic blocks ("cores").

1.6 Parallel Standards

STIL was not defined in a vacuum. There were quite a number of existing formats available. The problem with each existing format was two-fold: (1) the format was only accepted by a limited number of users, and (2) each format generally had limitations that other users found unacceptable. As available (and provided to the Working Group), these other formats were reviewed by the Working Group and considered for their strengths. The positive aspects of these efforts were then incorporated into the working definition.

This book references other efforts where appropriate, to compare and contrast those efforts with STIL. Some of these efforts are listed in the Bibliography section of this book.

Most formats proposed came from developments in the ASIC industry. These include formats from TI, IBM, and Motorola (at the time, ASIC gate-array division). LTX, as a supplier of test equipment, provided constructs of the enVision system for the effort. The IEEE Standards Association even had a parallel standardization effort with a very similar scope. This is listed as citation [2] in the Bibliography.

One particular definition was offered by a company called, at the time, Test System Strategies, Inc., or TSSI. While the company has undergone several name changes, the format they defined and continue to support is titled the Waveform Generation Language, or WGL (and pronounced "wiggle"). This is citation [3] in the Bibliography. WGL has been used both internally and externally by many companies and in several environments, and for people familiar with WGL constructs this book will compare and contrast WGL and STIL at appropriate times. For those not familiar with WGL these sections are not critical to the understanding of STIL, but will still provide an additional perspective on some of the language construct choices that were made in STIL.

CHAPTER 2

STIL Test

Standard Test Interface Language

This chapter presents an overview of the Test environment as it applies to STIL.

This entire chapter provides background on the primary application environment that STIL was designed to support. While references are made to how STIL addresses some of the topics here, this chapter does not present STIL language constructs.

2.1 Test Stresses

STIL is defined, as part of it's scope, to be a language to support Integrated Circuit testing, which establishes a set of testing requirements separate from other types of test environments. But even in IC testing there is a continuum of IC design styles, from full-custom to gate-array, with different testing issues. And even within these areas identifying test requirements can be a fuzzy operation. Some of the general test issues for different IC design styles are outlined in Figure 2.1. The next two sections discuss the environments at each end of the continuum, and how they affected the definition of STIL.

2.1.1 ASIC Test

ASIC Test environments contributed quite a body of formats in support of test transportation, due to the needs of the ASIC design flow.

ASIC Test has a particular focus with the issue of test transportation because of a combination of factors: generally small production runs and a requirement to support under-utilized and varied test equipment on the test floor. This places stress on a test flow that requires both first pass success capability and maximal portability. ASIC test environments respond to these stresses with a rigorous test flow integrated into the design process, a self-contained test language, and robust validation strategies of the test information. In addition to the STIL goal of test transportation being key to support this environment, the ability to support a single standard for the test interface is a benefit particularly with EDA tool suppliers faced with the previous varieties of test languages developed to support ASIC tool sets.

In order to facilitate test transportation between ATE equipment, the timing of test events in an ASIC test is often defined in a generic context, with restricted waveforms and broad timing, and identified to operate over a selection of ATE equipment. ATE-specific capability is not desired in this environment.

2.1.2 Microprocessor Test

Microprocessor test can be considered to be at the opposite end of the IC test spectrum from ASIC test, with a different set of test stresses.

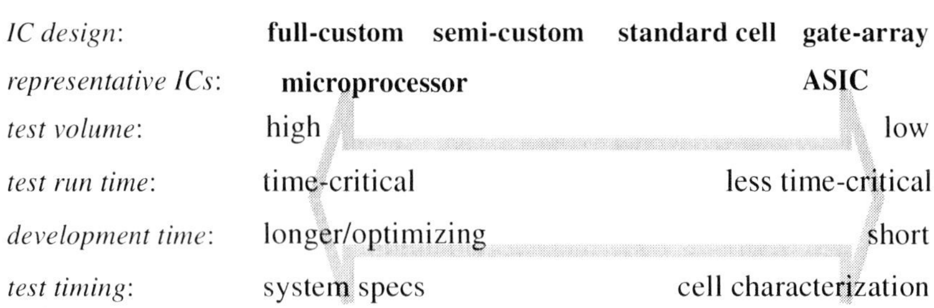

FIGURE 2.1 The IC Design Continuum and Test-Related Stresses

Microprocessor production may be very high volume, and test performance is a critical factor. Reducing the test time by very small amounts can reduce overall cost of manufacturing by dramatic amounts, impacting not only the per-device cost, but the overall the number of testers required, and the associated test floor space, maintenance costs, and all other support issues. Therefore the tests are highly optimized, and often specifically optimized for a particular tester (and configuration of that tester). The tests do not need to be portable because they will not be run on different test environments. And test development time is often much longer because of the necessity to optimize the test.

One of the requirements for microprocessor test is to validate that the device operates within specified system parameters. Often the microprocessor buses have a system specification defining the bus behavior, and even more often the overall microprocessor is designed against some type of a system specification. This is different than the ASIC design, and places requirements for additional capability in STIL for timing representation in the language. Rather than being defined from the bottom-up, the timing is specified top-down, often with system performance parameters that identify the minimum allowed timing, the maximum supported timing, and the expected, or typical timing of the system that contains this IC. STIL timing specifications were derived primarily to address these requirements.

2.2 Test Strategies

The sole purpose of test is to validate all specified constraints of a design, against an instance of that design, die or packaged component. Very generally, this comes down to two types of validation: (1) validation of the specifications of a design, generally against the ports of that design and the parametric capabilities of those ports (which are carefully specified by the vendor to support integration of the design into a system), and (2) some amount of validation of the design interconnections and appropriate or expected functional operation of the design.

TABLE 2.1 Some Types of Tests

Name	Purpose
Opens/ Shorts	Checks connectivity of the DUT to the ATE as well as fundamental operation of the device pins. First test performed.
Functional	Validates the device functional operation in it's primary operating, or mission, mode.
Structural	Validates the device structural interconnections and correct operation of all the lowest level components of the design.
Delay	Checks for timing related issues.
I_{DDq} / I_{SSq}	Measures quiescent current on the ground or power rails of the design. Detects defects that cause current leakage when the CMOS circuitry is in a static state.
Parametric	Validates Input and Output (IO) specifications. Most commonly associated with top level Signals testing, but may also test general operating (whole device) requirements as well.
Threshold	Validates IO and general operating specifications under functional operation, typically executing a functional test at specification limit values (that is, at one threshold of the operating limits).

Validation of the device operation is important to identify properly operating devices, and to eliminate non-functional devices. The trick in cost-effective testing is to identify properly functional devices as quickly as possible, and to eliminate the bad devices quickly as well from further testing.

To gauge the effectiveness of the test information, test development flows rely on representational models of device behavior, and perhaps more importantly, improper behavior to be detected.

2.2.1 Fault Models

Physical defects at the device level do not allow a direct mathematical treatment of testing and diagnosis. Therefore device defects are modeled using constructs that do lend themselves to evaluation. Fault models represent an abstraction on the types of failures that can happen in device circuits. They provide a mechanism to define a metric that can be used to validate test effectiveness, or to assist in selecting test sequences that provide overall test benefit. But always remember: fault models are an abstraction.

There are different types of fault models that represent particular types of faulty behaviors, and tests are developed in support of specific models. The following sections identify some of the testing strategies and the fault models commonly associated with that style of test.

2.2.2 Functional and Structural Testing

Functional testing refers to testing the device as it is meant to be used in the system. Functional test vectors are sometimes easily obtained, and sometimes are quite difficult to extract from the system-level environment. Sometimes these vectors can be used directly in the test environment, preserving the timing used in the system, and sometimes these vectors need to be modified to be compatible with the test environment. When the timing can be used in the test environment, the functional test is often referred to as an "at-speed" test. The biggest question when using functional vectors is: how effective are these vectors? How many of the device constructs and interconnects are being exercised by these vectors?

Structural testing is a test methodology that relies on additional design circuitry to support the capability for efficient test generation, often using completely automated tools to create the test vectors for the design. There are several design principles to support automated test generation, two in current use are scan-based and Built-In Self Test, or BIST. The elements of scan, and the support constructs in STIL for scan, are presented in Chapter 12, STIL Scan. Structural testing validates the proper operation of all the design cells, and all interconnects, but does not exercise the device in a functional mode to accomplish this validation.

Both structural and functional testing generally use the Single Stuck-At fault model to evaluate the effectiveness of the test vectors.

2.2.3 Single Stuck-At Fault Model

Stuck-At fault models represent device behavior when nodes are held at a constant value, hence the term stuck-at. There are several types of stuck behavior, but the most common is the Single Stuck-At Fault. This stuck-at model use constant logic values, either a logic high or a logic low as the stuck condition, resulting in a pair of faults on each node, a stuck-at-one and stuck-at-zero condition for each node. Nodes fixed at a single logic level, for instance nodes fixed at a logic high state, cannot detect the fault of that state, in this case a stuck-at-one fault, because they cannot be set to low and then have that low state detected and propagated out. It is necessary to set the node to the opposite state and detect the presence of that state in the output, to detect that stuck-at fault.

Single stuck-at models are commonly used as the metric of structural and functional test validity. The total number of faults in a design is fundamentally the number of nets or connections in the design, doubled for each of the primary stuck-at conditions (stuck-at-one and stuck-at-zero). The coverage metric is derived from the ratio of the faults detected over the total number of faults in the design.

In order to detect a stuck-at condition, an output response must be identified that would indicate if the particular node was actually stuck in the particular condition. The effort to detect a stuck condition requires a path sensitization to that failing node, and a propagation of that faulty state to an output to be detected. Both of these operations are key to the algorithms and processing operations of automatic test pattern generation tools. Figure 2.2 presents a simple example of the path sensitization process.

2.2.4 Bridging Faults Model

A permutation on the single stuck-at fault model is the bridging fault model. Bridging faults are similar to stuck-at models, but represent a condition where one node is stuck-at the logic state of another node. Bridging faults are detected when the two nodes are set to opposite states and the current state of the node is detected by a test response. Whereas, in the case of stuck-at faults, the total fault set can be computed from the netlist, there are far too many possible combinations of bridging faults. Potential bridging faults are usually determined by examination of the trace proximity of the net layout information to identify a set of candidate bridging faults.

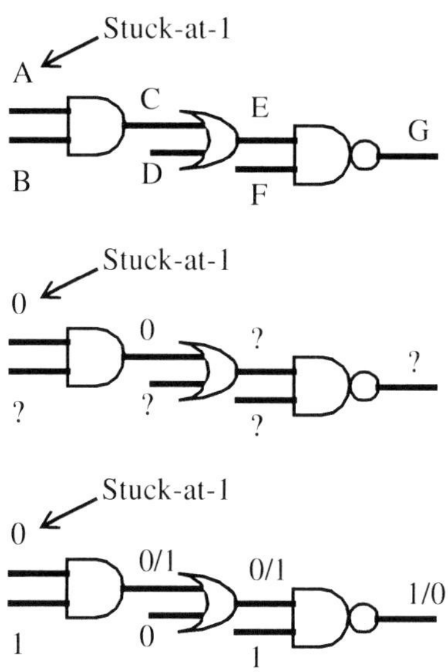

FIGURE 2.2 Single-Stuck-At Path Sensitization

2.2.5 I_{DDq} Testing

I_{DDq} testing is very different than stuck-fault testing. I_{DDq} tests are based on the principle that CMOS devices have very low quiescent current, and many types of CMOS defects manifest as elevated quiescent current through device leakage paths. I_{DDq} faults are observed by measuring the steady state device current through the V_{DD} pin of a CMOS circuit. This current measurement is made after an I_{DDq} test vector has been applied and any transient currents within the design have dissipated.

The current measurement in a CMOS device in the quiescent state is very small, and perturbations in this current caused by faulty devices may also be small. This issue affects the dynamics of the test measurement itself, which may take a long time (long in terms of overall test time) to perform the accurate measurement necessary. This is the negative side of I_{DDq} testing; they may be expensive to perform at test. Test strategies such as on-chip current detectors or special test hardware placed physically adjacent to the DUT can reduce test time at thc cxpcnsc of additional test circuitry.

But I_{DDq} tests have several advantages. In addition to testing device leakage paths, I_{DDq} vectors can test for bridging paths on cell interconnects in the device by enforcing a detect condition as the opposite state between the two bridged cells.

Another advantage of I_{DDq} testing is that the failure condition — elevated current — is detectable based only on the current device state. These states do not need to be propagated to output nodes to be detected, as a current measurement with the nodes in that state is sufficient to test for those faults. This eliminates the path sensitization process when generating tests for I_{DDq} tests.

I_{DDq} tests may be generated using several different fault models. Three of these models are presented next.

The Quietest fault model was developed directly from consideration of transistor structures. It identifies six potential faults per transistor, observable from input and output conditions on the transistor. The disadvantage of the Quietest model is that the model representation is limited to two states and it may be difficult to generate models to represent complex cells. EDA tools do not use this model in practice.

The Toggle fault model represents cell-level inputs and outputs as stuck high or stuck low, but eliminates the path sensitization and fault propagation processes from the Single Stuck-At model. This model accurately captures device state at the cell level, but may not present a thorough internal test for larger cells in the design. However, toggle detect (and bridging detect) can be extracted from many digital simulation tools without a fault simulation context, and I_{DDq} fault grades from this model can often be accomplished using the standard simulation tools.

The Pseudo Stuck-At model addresses the weakness of the Toggle fault model. As with the Toggle model, this model presents cell-level inputs and outputs as the nodes for fault detection, but requires input fault effects to be propagated to the cell outputs before they are detected. This makes this model equivalent to the Single Stuck-At model at the cell level. This supports a level of internal cell coverage not accomplished with the Toggle model, providing a more complete set of tests. Another implication of this behavior is that faults on a multiple fanout net are not equivalent as they propagate into other cells. The limitation of this model is that the fault propagation through the cell may not cover complex or sequential cell behavior robustly.

2.2.6 Delay Tests

Functional and Structural test strategies validate device behavior, but may not robustly validate the device timing. Delay testing may be necessary to detect timing problems in the design; there may be manufacturing problems which prevent the device from operating at-speed but passes structural testing. Many defects at 0.18μ gate dimension and below are timing related. Additional delay from these defects is usually small and often less than the cycle time used for functional or structural testing.

Delay Tests patterns are generated and measured against a fault model for timing related defects. Delay fault models use a two vector capture sequence and require both a launch event and a capture event with appropriate timing expected between those events. There are two fault models commonly used for delay testing which are transition fault and path delay fault.

Delay fault types are common across both fault models, and consist of two types: a Slow-To-Rise and a Slow-To-Fall fault.

2.2.7 Transition Fault Model

A transition delay fault is a defect that would cause combinational logic along a valid signal path to transition either too slowly or too quickly for the circuit to function correctly at-speed.

Transition delay fault tests use a fault model where the defect is assumed to occur at a single node of a gate and be large enough that the specific propagation path to an observable capture point allows that defect to be captured. Transition delay fault sites are usually the same as stuck-at fault sites. Figure 2.3 presents one example of this type of fault.

Transition Delay fault models target *point defects* that effect the timing. Some point defects include metal scratches, dark or missing vias, poly cracks and metal or poly bridges on the device.

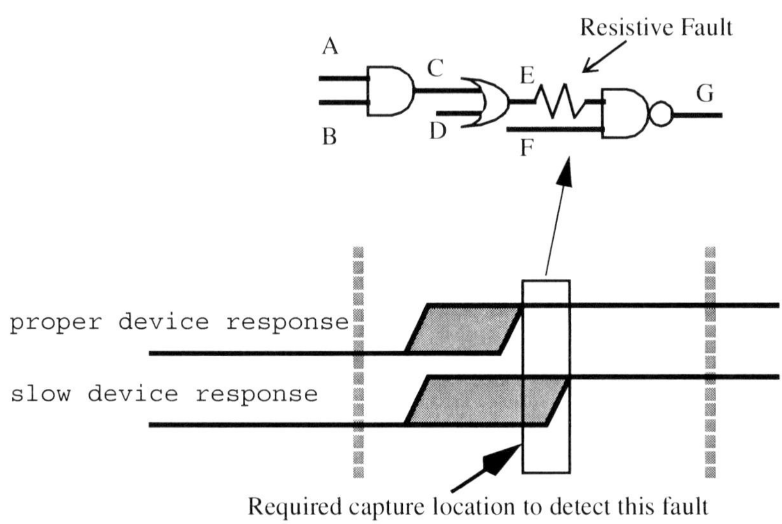

FIGURE 2.3 Example Transition Delay Fault

2.2.8 Path Delay Fault Model

Path Delay Fault Tests use a fault model where the defect(s) may occur at one or more points along a path, usually a critical timing path. For path delay fault tests, the propagation path is fully specified to ensure that the complete path is tested. The target paths for path delay fault tests are usually provided by a static timing analyzer tool.

Path Delay targets *distributed defects* that effect the process timing that do not allow the chip to work at maximum rated speed. These distributed defects include mask alignment, ion implant levels, oxide thickness, and metal widths or metal thickness.

2.2.9 Test Metrics

The measurement or metric of test coverage, to identify the goodness of a test, is based on the fault model used to develop or fault-grade the test. As we identified above, the basic metric for a test is the number of faults detected over the total number of that type of fault in the design.

Wide use of the single stuck-at model in the industry has indicated that high single stuck-at fault coverage correlates to a high level of defect coverage. As IC technologies evolve, however, the defect distribution changes and additional test strategies are continually defined to cover these issues. For instance, small design geometries have increasing sensitivity with timing related issues, putting additional stress on delay test strategies to generate tests that effectively test these designs.

It is important to remember that tests are generated in the context of a particular (assumed) fault context. Fault simulation, if available, may always be used to evaluate the effectiveness of a test for its ability to detect the faults for a specific type of fault model. But fault simulation effectiveness is directly related to the accuracy of the fault model. And the effectiveness of the test in this environment is tied to how appropriate this test is for detecting this type of fault.

2.2.10 Fault Diagnosis

When devices fail during test, the test failures may be used as a signature to identify a defect or a root cause of the failure. Tracing back, via the failing test signature, is called failure diagnosis and typically involves processing more information about the overall test than is present in the test itself. Failure identification typically relies on these same fault models as well.

Fault diagnosis is heuristic in nature and makes assumptions about the types of faults present in the design, to detect those faults. It also makes assumption about the test results themselves (and how the test was run). For instance, not only at what point in a test was reported failing, but that everywhere else — including after this test — did not fail. This implies that executing tests for the purpose of effective diagnosis is different than running the tests in a normal production mode, which typically stops the testing at the first design failure.

The test itself may be constructed differently for diagnosis or more complete testing operations. An example of this is the use of "N-detect" test generation for structural tests. By default, structural test generators often target a single primary detection of a particular fault. But sometimes detecting that same fault multiple times (hence the term "N-detect", where "N" is the number of times the fault is detected) can result in more information that can isolate additional types of failures and result in more thorough device testing.

2.3 Test Behaviors

There are certain characteristics of a test program that are critical to understand. These characteristics apply to ATE architectures and to the constructs of STIL itself. The next sections identify these characteristics and how they pertain to both the ATE architecture and to STIL.

2.3.1 Signals

Automatic Test Equipment connects to the Device Under Test at the Signals. Signals on ATE have very specific capabilities, in particular drive capabilities (voltage, current, slew characteristics), measure capabilities (detect levels, timing), and the ability to dynamically change during the test. To test a device, the ATE needs to be able to operate the DUT with enough capability to stress the device — the dynamic capabilities of the ATE needs to be sufficient to stress the device sufficiently to validate that the device will operate properly in a real application.

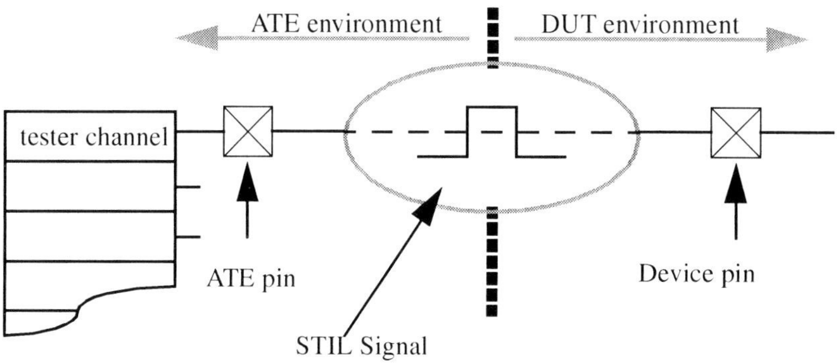

FIGURE 2.4 STIL Signals Between the Device and Test Domains

All Signals are not equal. Some Signals just provide power and ground connections to the DUT. The dynamic requirements of these connections are pretty limited, but the current (driving) capacity is often an issue. In some situations, power and ground connections may be special connections to the DUT to insure adequate current capacity. Other Signals may supply input stimulus to the device, while others may be generating output response from the device only. Even though input Signals only supply stimulus, and output Signals may only provide response, both types of Signals are very often connected to an ATE Signal that can perform functions necessary for both types of behaviors.

And then there are the Signals from the DUT that do behave with both Input signal behaviors, and Output responses. These Signals are called *bidirectionals*. In most situations, all communicating Signals at the ATE are bidirectionals, even if the DUT only requires a part of the available functionality of the Signal.

Bidirectional device signals must be tested carefully. Applying stimulus from an ATE at the same time that the DUT is driving the Signal will result in drive conflicts. The stronger driver, if driving in contention for long enough, will burn-out the weaker device or shorten the life span of operation from this stress. Generally the pin electronics of the ATE environment have more capacity and drive capability than the drivers of the DUT, and generally if prolonged contention occurs during test it will be the DUT that will suffer.

A representation of a full bidirectional configuration for both the ATE and the DUT is shown in Figure 2.5. This figure identifies the 6 pieces of information that are present in this environment — the drive, enable, and measure values on the tester side and the corresponding information on the device side.

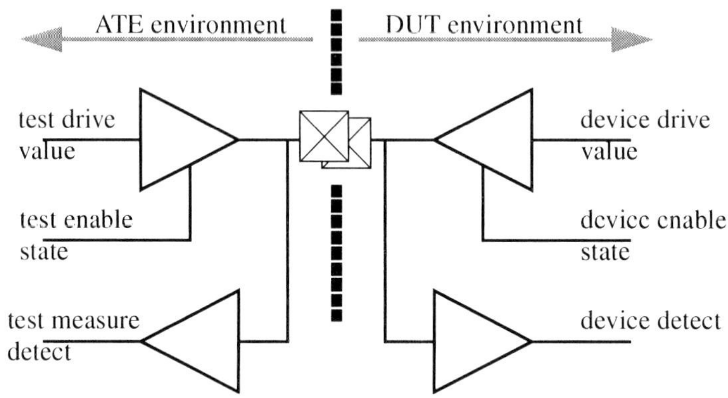

FIGURE 2.5 Schematic Diagram of a Fully Bidirectional Signal at Test

The test requirements to support proper bidirectional operation will be presented later. The constructs presented in Figure 2.5 are important to remember in these discussions.

Figure 2.5 presents a small part of the ATE circuitry present on most Signals. Most ATE supports additional devices and circuits necessary to completely test a device. Some of these additional circuits are presented in Figure 2.6.

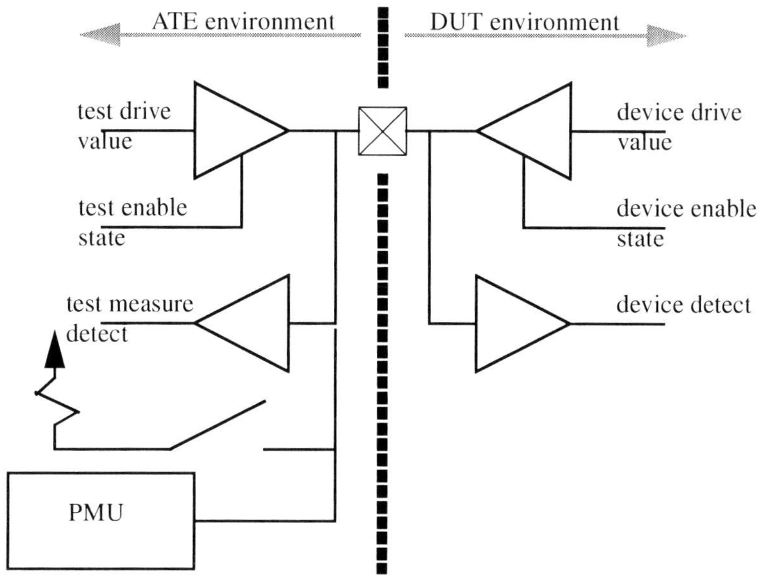

FIGURE 2.6 Some Additional Test Circuitry

2.3.2 Stimulus and Response

Device test consists of two fundamental sets of information: the stimulus necessary to operate the device, and the capability to observe the device response and validate that it is appropriate. A properly operating device will respond as expected to a set of stimulus; an improperly operating device will be detected as not responding as expected. When an improperly operating device is detected as not responding as expected, it is said to *fail* the test.

2.3.3 Periodicity

ATE has its own set of limited capabilities and behaviors. In general, while the ATE has a certain amount of timing flexibility it cannot generate a completely free-form set of stimulus events, or as we will present next, output response checks.

The limitations on timing flexibility in the ATE environment requires consideration as the test is being created. To support generating stimulus appropriate to the ATE environment, timing events in the stimulus are constrained to regular, repeating events when possible. For examples, a clock pulse necessary to operate the device, may be timed to occur at regular spacings between the edges of the clock. These regular spacings are further separated into repeating time intervals. The entire process of taking a stream of events and compartmentalizing the events, particularly into uniform time sequences or periods, is called *cyclization.*

ATE environments can support cyclized data with very compact representation, which in turn facilitates holding large amounts of test data to send to the device. Also, by minimizing the amount of unique time values, the ATE environment can be optimized to present the defined timing with very high precision and repeatability.

The repeating event cycles that exercise the device (and validate the response) are called *Vectors* in STIL (and also in most ATE environments).

2.3.4 Output Response Constraints

Device output signal response is a function of the device input signal stimulus and the specific characteristics of the device itself. Device response occurs in real time, and is not naturally partitioned into the cycles presented above.

However, again due mostly to ATE considerations, the output response information present in the test program must be reduced to something compact. Typically this becomes a single timing reference (and remember the discussion above about timing restrictions) and a specific expected value for the device response at this time for each test cycle. This is called a *test strobe.*

The placement of the test strobe is critical for test operation. First of all, for effective testing, the strobe should validate some device response or design fault uniquely. Secondly, the strobe needs to be placed at a time appropriate for identifying a functioning device. Guardbanding the strobe placement is necessary to create a robust test. Guardbanding constructs are presented in "Guardbanding" on page 30.

There are two primary types of test strobes, identified by the width of the region being validated. *Edge strobes* occur at a single time within a period, and validate the device state only for that instant in time. *Window strobes* validate the device response over a region of time. During this region the state of the output being strobed cannot change. Window strobes provide a level of additional test coverage by validating the output is stable over this region of time.

2.4 Test Induced Faults

Testing a device is not without the potential of introducing a class of faults itself. Certain aspects of test behaviors can result in introducing faults into the design, which may result in device failures during testing that are a result of the test operation itself.

One goal of device testing is to validate device operation under an environment that is as close as possible to the normal operating behavior of the device. This is not always possible, especially for efficient testing. Scan-based testing, for example, provides direct access to the sequential elements of a design, which provides a shortcut to establishing the state of a design necessary to test the internal connections of the design. However scan-based testing sacrifices normal device operation to achieve this shortcut.

One potential impact of this operation can be a level of higher internal activity in the design than would be expected in normal functional operation. This can result in several detrimental side effects, including higher operating temperatures during test and increased opportunity for ground-bounce problems.

2.4.1 Packaging Effects

Not exactly a test-induced fault, but test behavior can be sensitive to the type of package being tested. The die testing environment has different characteristics than a fully-packaged test environment. Packaged die often have a ground plane as part of the package, and current-induced situations are mitigated with the presence of this ground plane. Die probes may have a higher inductance than a package test socket, due to the longer traces of very narrow wires. These and other considerations may affect how test is performed between die and package operations.

2.4.2 Simultaneous Switching Outputs

Signals at the interface to the design are special. For instance they contain special circuitry to protect the device from static electricity. Output components must be physically larger to drive more current and different voltage levels off-chip. These larger components contribute to much of the device power consumption, particularly when they switch values.

Structured test generation using IEEE Std. 1149.1 (JTAG) constructs in the IO ring of a design can in some situations contribute to test-induced problems. One situation caused by an attempt to minimize the number of tests is to transition all the output states at the same time, to different values. This can result in ground-bounce problems and may result is identifying limitations on the number of Simultaneously Switching Outputs (SSO) constraints necessary to avoid ground-bounce at test.

2.4.3 Guardbanding

The placement of output strobe timing adjacent to device output transitions can cause unintentional test failures. These failures can be caused by a combination of factors, for example, slower (or faster) device response, or changes in the timing of the strobe placement by the tester.

To eliminate failures at test due to these situations, test strobes need to be placed at a location safely away from the actual device transition. This safety zone is the *guardband*. Guardbands are cumulative; they are the sum of the variation of allowable device response time plus the variation in the ATE strobe placement timing, plus other factors that need to be protected for, for example variations in capacitance in the test configuration that affect the measurement time.

2.5 *Other Applications of STIL*

So far we have talked about the issues of the Test environment, which STIL was primarily scoped to support. However, there are other application environments where STIL is used and may even be required.

One such application is as an input language to control design rules checking of scan based designs. The motivation for using STIL in this context is to minimize the number of different formats needed to operate the tool. The purpose of the design rules are to facilitate ATPG generation of test vectors. By using STIL as the input to this process, the output test vectors from the ATPG process can use the same STIL information during the test generation process, and maintain this information as part of the test data. This then allows for an iterative loop around the ATPG process, to refine the rules and to support readback of the test data for subsequent ATPG runs and diagnosis operations. Figure 2.7 presents this advantage. The iterative loop is shown in this figure by the arrow going around the outside and connecting the two sets of STIL information.

Another context of application for STIL is to apply this information under event-based digital simulation, such as that defined by citation [4] of the Bibliography. Event-based simulation is often used in design environments to validate design behavior, and the ability to use STIL in this context allows the test formatted information to be applied as stimulus and response data in the simulation context. This can allow for early debug of test programs, and examination of test failures off-line from ATE hardware. However, the primary focus of STIL as a transport vehicle from EDA to ATE is seen to limit to overall utility of the language in a design context, and is one of the driving factors for the first extension effort listed in Table 1.1, "STIL Subprojects," on page 10.

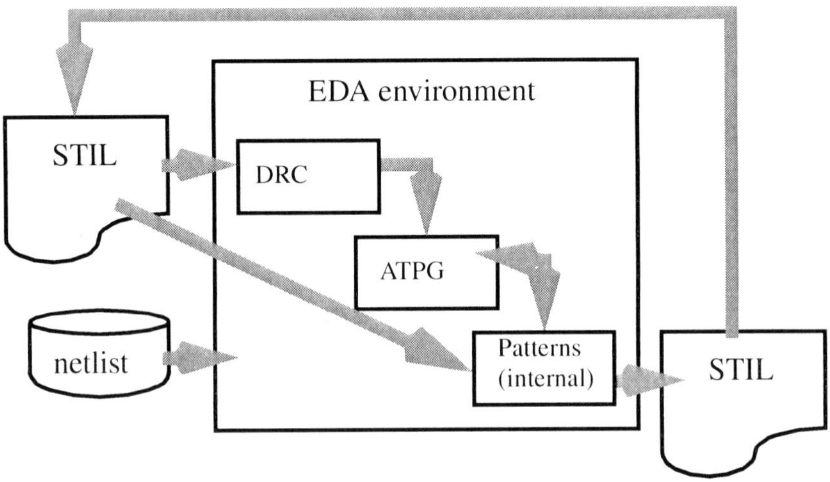

FIGURE 2.7 One Benefit of Using STIL in EDA

CHAPTER 3

Signals

Standard
Test Interface
Language

STIL files start with a single statement at the top of the file indicating the file contains STIL information (this statement is presented in Section 6.4). But the first block of information pertinent to the test environment is the Signals block.

3.1 Signals

STIL takes a device-centric view of the test process. The reason for this is that a STIL test program is expected to be transportable across a variety of test platforms. The only intransient aspect of this environment is the device to be tested. Figure 3.1 visualizes this perspective.

All stimulus and response defined as part of the test program is specified relative the ports of the design. These ports are called Signals in STIL terminology. The Signal block in a STIL test program contains all the ports of the design, and may include additional ports that are part of the test operation, necessary to support applying test stimulus and validating appropriate design response.

STIL Signals are referenced from the perspective of the DUT

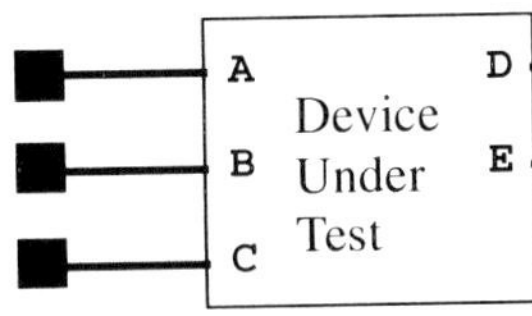

FIGURE 3.1 STIL takes a device-centric view of test

Signal names are limited to the acceptable definitions of a STIL user name as identified below. This book will refer to all identifiers and names defined by the user as user names; the STIL Standard calls these user-defined names.

The limitations on the contents of user names is a concession to the test environment. The test environment is often limited on the way names that can be represented at test. This is conflict with a device-centric view where device pin names are often quite elaborate and may contain a variety of special characters. Consider for example, conventions for indicating inverted signal values: !name, name~, name_NOT, INV_name, etc. The variety of conventions identifies the lack of a single convention. Part of the benefit of STIL, and standards in general, is to reduce the options, but still allow for *defining* a single convention to be applied in any particular environment.

The STIL mechanism to allow complex design names is to enclose the names within double quotes. Double quoted names may contain any printable ascii character, including blanks and tabs, except, of course, the double quote character itself.

The primary goal of STIL is to apply this information at test. A test program containing only device-oriented names will not run at test without some more information — for example, the relationship of test connections to each of these Signals. This issue of providing a correspondence for each name with a test connection would appear to be unaddressed in the STIL Standard. It was actually the point of several discussions. A common mechanism was not established as part of the discussions and is not defined as part of the 1450.0

standard. However, it was identified that all test flows have mechanisms for establishing the test connectivity, and the goal of the standard became to provide a mechanism for each flow. These mechanisms are outlined in Chapter 17, Name Mapping.

The next couple of sections cover the basic characteristics of user names in STIL.

3.1.1 User Names Characteristics

User names (which include Signal names) must begin with a alphabetic character or an underscore, and may contain any sequence of alphanumeric characters and the underscore as long as that sequence is not another reserved STIL keyword. If any other character is present then the entire name must be enclosed in double quotes. User names are case sensitive; names with different cases are different names.

Double quotes allow a user name to contain any character, which means that the only character that should (normally) terminate a string that starts with a double quote is another double quote. Double quotes can also be used to differentiate a user name from a matching STIL keyword. Enclosing a name in double quotes separates it from a STIL keyword as STIL keywords are never quoted.

STIL Conventions

It is recommended to uniformly quote all Signal names in a STIL test program in order to maintain design data through a STIL flow. Remember that, once quoted, all references to that Signal must be made with the quotes; in fact if both the quoted and unquoted forms of a name exist in a STIL program they reference different Signals. By uniformly quoting *all* User names in the STIL file, confusion about what names might be quoted and what are not is also eliminated.

The double quote character cannot be part of a Signal name, except as the first and last characters of a quoted name. While some environments will accept a double quote as part of a name, that name must be modified to eliminate the double quote character for a STIL flow.

It is not recommended to enclose carriage returns as part of a user name, although technically the only termination for a double quoted name is the closing double quote. Any carriage return contained within a double quote does become part of that reference, and any use of that name must be identical, including the carriage return at the same position. To partition really long names across multiple lines and not include the carriage return as part of the name, use the concatenation operator presented in section 3.1.2.

3.1.2 Really Long User Names and Concatenation

A single token is a consecutive sequence of non-whitespace characters; whitespace in STIL is a blank, tab, and carriage return. Also, most punctuation characters will terminate tokens. Specific characters found in STIL files are listed in Table 3.1. Many of these characters are reserved to specific contexts and may not normally terminate tokens except in those contexts. But carriage returns, braces and semicolons do commonly terminate tokens in a STIL fie.

TABLE 3.1 **Characters that Terminate STIL Tokens**

Character	Name	STIL Application
;	semicolon	terminates STIL statements
{ }	braces	terminates STIL blocks
[ ]	brackets	specifies signal arrays
:	colon	terminates labels
/	forward slash	divisor operator, separates state characters
@	at-sign	time-marks
,	comma	user function lists
.	period	concatenation, hierarchical timing names, and spec variable type referencing
!	exclamation point	scancell inversions
#	pound sign	argument substitution
%	percent	argument substitution

TABLE 3.1 Characters that Terminate STIL Tokens

Character	Name	STIL Application
\	backslash	starts vector flags
( )	parenthesis	argument lists in procedures, macros and functions
// /* */	comment lines or blocks	comments

STIL limits the length of a single token to 1024 characters. There are circumstances where user defined names may exceed this length (typically top level Signal names are not this long, but ScanCell names that will be presented later may be longer). To support an unlimited name length, STIL uses a period ('.') between two tokens, to indicate that these tokens are to be concatenated together as a single name.

Any user-defined name may use the concatenation operator to combine shorter tokens into one longer name; tokens do not have to be 1024 characters long before the period may be used.

The period operator is applied in another context as well, in the Timing section. To avoid STIL ambiguities it is recommended to minimize (or avoid entirely) mixing the application of the concatenation and hierarchical references in the Timing block. This is covered in detail in Chapter 10, Advanced Timing.

STIL Conventions

When referencing a concatenated user name, that reference should be processed as if concatenation operator was not present. A reference containing the concatenation operator is the same as a reference without that operator. For example, a . b . c is the same as abc.

It is recommended that all applications processing STIL data maintain some awareness of the concatenation operator when present during processing, in order to regenerate the same name when writing STIL output. If your application does not write STIL out then the concatenation operator may be discarded for the purposes of processing.

Concatenation may be used between unquoted and quoted strings. If any substring is quoted (whether required or not), the entire user name should be considered as if quoted. For example, a . "b" . c is the same name as "abc".

Generators of STIL data should minimize the use of the concatenation operator, and ideally use it consistently when it is used, to minimize confusion with multiple forms of the same reference.

3.1.3 Types of Signals

Each Signal declaration has a type associated with it. There are five types of Signals in STIL, listed in Table 3.2.

TABLE 3.2 Signal Types in STIL

Name	Behavior
`In`	Signals with stimulus information only from the test.
`Out`	Signals with response data only defined in the test.
`InOut`	Signals that contain both stimulus and response data in the test. Also known as bidirectionals, these signals often require special handling in the test program to change direction from input to output or vice-versa.
`Pseudo`	Signals that are associated with the test operation but may not be part of the design itself.
`Supply`	Signals used to power the device and do not require stimulus and response data themselves.

While these types indicate an expectation on the type of stimulus or response data present, there are no requirements in the STIL environment that the data present on any one of these signals be constrained. However, data that is not supported by this type of Signal may be ignored (or better, generate a warning that the data will not be used) and not applied in the particular context if that context cannot support that type of data for that type of Signal. For example, it is not incorrect for an input Signal to have both input and output (response) data present. The output data may serve as validation of the input state application; it typically would not be additional response behavior.

3.2 Signal Arrays

STIL reserves square brackets at the end of a Signal name to represent an arrayed Signal reference. The intention of arrayed references is to serve as a shorthand declaration of a group of Signals with common attributes, typically representing a bus or collection of associated Signals in the design. The brackets surround a range specification which is a contiguous inclusive set of numbers between a starting value and an ending value, both of which are integer values. These two integer values are separated by two period characters, as a sort of a shorthand ellipsis operator. For example:

```
(3.1) "ibus"[2..33]
```

When present, the square brackets (and the contents) become part of the Signal name or reference in the same way that quotes (if present) are part of that name. There is no shorthand reference to all Signals in an arrayed declaration by referencing the name without brackets; the Signal names **`"bus"`** and **`"bus"[0..15]`** are references to two different sets of Signals.

Quotes, if present around a Signal name, enclose the text part of the name and do not enclose the array part of the name. Enclosing the brackets inside the quoted part of a name results in declaring a single Signal of that entire name.

STIL Conventions

Arrayed Signal declarations in STIL should **not** be equated with vector constructs in netlisting languages. While STIL arrays do have similarities with netlisting constructs, they support in general a subset of behaviors associated with netlist constructs. However, it may be a useful convention to pass netlist vectors in a design as a STIL arrayed Signal declaration. Another mechanism of maintaining this data is to define the collection of Signals as a SignalGroup, which is covered in the next section.

Do not put a blank (or any whitespace) between the text and the opening bracket of an arrayed Signal reference. You may have whitespace inside the brackets of an arrayed Signal reference, around the integer value, if desired. But the brackets are considered part of the Signal name, and Signal names cannot contain whitespace (unless the name is enclosed in quotes). The leading bracket needs to be part of the Signal name — the remaining part of the array reference is associated with the opening bracket as part of the array index.

Do not put a blank (or any whitespace) between the two periods in the array range operator. Again, whitespace may be placed around the integer values and this operator, but to avoid potential parsing confusion the two dots of the array operator should never be separated.

3.3 SignalGroups

In addition to arrays, Signals declared separately may be defined and referenced as a group. Groups are additional references that do not declare new signals, they only reference collections of existing ones, or may serve to rename a single Signal to a new name. Groups are declared in the SignalGroups block.

SignalGroups are ordered lists of Signals, and the order of the Signals in a Group is critical to proper STIL processing. Whenever a SignalGroup is used, the first Signal in that Group receives (generally) the first value of an assignment or other application of that Group, and the last Signal receives the last value. This is shown in Figure 3.2. This process always occurs aligned to the Signal order, and should not be confused with the constructs of LSB (Least

Significant Bit) and MSB (Most Significant Bit), attributes that may be associated with Groups which are presented later.

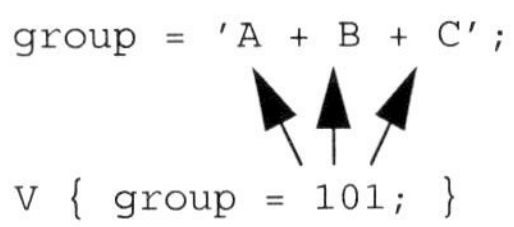

FIGURE 3.2 Example of SignalGroup Data Ordering

There are no restrictions on the types of Signals that may be grouped together. It is expected that In and Out Signals may be grouped together. In fact, it is expected that Groups will contain a mixture of Signal types identified in Table 3.2.

Because a SignalGroup is an ordered list, there are only two primary operations supported to create these lists — insertion, represented with a "+" operator, and deletion, represented with a "-" operator. The order of the Signals in a SignalGroup is the order that the Signals are referenced in the SignalGroup expression.

Arrayed Signals may also be referenced in a SignalGroup expression with a shorthand format, using the STIL ellipsis operator ("..") to represent a contiguous range of Signals. The ellipsis operator is equivalent to a series of "+" operations on the sequence of individual signals. For example, "A"[1..2] is the same as '"A"[1] + "A"[2]'.

A SignalGroup expression is always delimited in STIL with the use of single quotation marks (" ' "). Single quotation marks are used to delimit several types of STIL expressions; basically it is a general STIL rule that any expression operators like "+" and "-" are always found inside expressions enclosed in single quotation marks. Note that a single Signal reference, without operators, may be referenced without single quotation marks.

Any particular Signal can appear only once in a SignalGroup. It is an error to have the same Signal present more than once in a SignalGroup, even as part of an evaluation process to define a Group. For this reason, parenthesis may also be used as part of the expression constructs, in particular to eliminate a Signal (or another Group) from a particular sub-expression before that expression is added to another expression that also contains that Signal. This is demonstrated in the next example.

```
Signals {
    A In; B Out; C In;
}
SignalGroups {
    E='A+B';
    F='A+C';
    G='E+F'; // Error: A is referenced twice
    G='E+(F-A)';// Correct: A is removed
                // from Group F first
}
```

In addition to the restriction on adding the same Signal to a existing Group, it is also an error to remove a Signal from a Group or sub-expression that did not contain that Signal.

STIL generally maintains a "define-before-use" requirement. For example, Signals and SignalGroups names need to be defined (or declared) before they can be referenced. In general this requirement facilitates a level of early data validation when the information is declared, particularly for ATE Environments that often have constraints on the amount of this information they support. In practice, many situations require more complete processing of the data before environment constraints can be accurately checked.

To a certain extent, one violation of this define-before-use requirement is the ability to organize a collection of Signals where they are used. This collection of Signals may be applied as a dynamic group, without declaration, in several STIL contexts, generally by using the single quotations around the group expression at a point where a Signal or SignalGroup name is supported. This is referred to as a “group-on-the-fly” in the STIL context. “Groups-on-the-fly” do not have a group name associated with them, and every reference to this collection requires repeating the complete organization of Signals. Therefore some of the advantages of defining SignalGroups with names is lost when groups are defined this way.

STIL Conventions

Consider using naming conventions on SignalGroup names to avoid conflict with the expected usage of a set of tests. For example, using underscores or numbers in the names will prevent conflict with STIL keywords (which by convention never contain underscores). Keyword conflicts are only a problem if a test program is integrated into a flow where multiple versions or extensions of STIL are present (see Chapter 15, User Extensions). Using a naming convention can also facilitate visual review of a STIL file by making it easy to recognize a SignalGroup name from the convention.

The STIL ellipsis operator can be used to reference any set of Signals declared with array indicators ("[..]") as long as the leading username is the same, even if all Signals were not declared together. For example:

```
(3.12)Signals {    Sig[0..10] In;
(3.13)             Sig[11..21] Out;}
(3.14)SignalGroup { allSigs = 'Sig[0..21]'; }
```

It is recommended to always use single quotation marks in SignalGroup expressions to provide for a uniform representation of expressions and facilitate reading of the STIL data. In particular, STIL allows a single Signal or Group reference to be made without the use of quotation marks; the primary requirement is that expressions containing "+" and "-" be enclosed in quotation marks. This *does* allow an arrayed Signal to be referenced without quotation marks if no additional Signals are inserted or deleted, for example:

```
(3.15)Signals { Sig[0..43] InOut; }
(3.16)SignalGroup { UpperBits=Sig[0..21]; }
```

While this is proper STIL, it is also not an error to place single quotation marks around these expressions either, and it is recommended to follow the practice of putting single quotation marks around all these expressions.

3.3.1 SignalGroup Block Names

The STIL Signals block previously presented can appear only once in a STIL test, and all Signals used in that test must be defined in that one block.

STIL SignalGroups are not so restricted. There may be multiple STIL SignalGroup blocks, following certain restrictions, and each block may define a separate set of SignalGroups.

STIL defines a robust hierarchy of SignalGroups blocks that may be part of a particular test context. The two following rules are generally applied in many STIL contexts:

- An "unnamed" STIL block is generally accessible from all contexts that use the information contained in that block. This is referred to as the "unnamed global block" or just the "global block" in the STIL specification. There can only be one unnamed block of any STIL type present in a STIL test (where "type" refers to the block name, for example "SignalGroups"). It is an error to have multiple unnamed blocks of the same block name present.
- A "named" STIL block is only accessible from contexts that expressly reference that named block. For STIL tests, these blocks are most commonly referenced from statements found in the PatternExec or PatternBurst blocks. This is presented in Chapter 6, Tying It All Together.

More information about naming STIL blocks can be found in "Named Blocks" on page 48, Section 3.4.3.

The next two sections are important to understand how Signals are resolved in SignalGroups, but is quite detailed in the resolution behavior.

3.3.2 SignalGroup Evaluation and Name Resolution

STIL SignalGroup definitions are fully resolved in each SignalGroup block using the following rules. These rules allow SignalGroup names to be fully resolved down to the individual Signals that constitute that group when the SignalGroup is declared; there is no run time or test context information necessary to resolve SignalGroups. Again (and I apologize for the forward-reference), these rules are differentiated on whether the SignalGroup block is named or not, and more information about naming blocks can be found in section Section 3.4.3.

- A SignalGroup name is resolved when it is declared and SignalGroups are resolved in the linear order (starting at the top of the block) as they are defined. A SignalGroup name may be defined multiple times in one SignalGroup, the last definition is the one used for that reference. However, if another SignalGroup references this name, the *current* definition of that name is applied when resolving this group. This is presented, hopefully more clearly, in the example following these rules.
- A SignalGroup name can be defined in only one named SignalGroup block across all named SignalGroup blocks that are used in one test context (referencing SignalGroup blocks to be applied in one test context is defined in Chapter 6, Tying It All Together).
- A SignalGroup name in a named SignalGroup block can replace (hide) a name defined in the unnamed SignalGroup block or can replace (hide) a Signal name. Once a name has been hidden it cannot be directly used for this hidden Signal again, however another SignalGroup may contain that original reference or be an alias for that original signal as long as it was defined before that original Signal or SignalGroup was hidden.
- A SignalGroup name defined in the unnamed SignalGroups block cannot replace (hide) a Signal name. Only names in named SignalGroup blocks can hide Signals. Because the unnamed SignalGroup block is globally accessible, if it was allowed to replace a Signal with another reference, the hidden Signal would never be accessible in this STIL test.
- A SignalGroup name reference (on the right side of the equals in a SignalGroup) is resolved against Signal and SignalGroup names in the following precedence: (1) if the name is defined as a SignalGroup name previously defined in this block, then that definition is used to resolve this reference, (2) if a global (unnamed) SignalGroup exists and the name is defined as a SignalGroup in this block, then that definition is used to resolve this reference, (3) the name is resolved from the Signals.

These rules identify a SignalGroup resolution process that is order dependent. The last definition of a name is the definition that is used for that name, however if that name was defined multiple times and was referenced between the different definitions, then that name may have had different values applied to other SignalGroups. This is important for contexts where test data is reused, for example with Core Test Language constructs. The following example demonstrates the importance of order dependent name resolution.

```
Signals {
    A1 In; A2 Out;
}
SignalGroups {
    A3 = 'A1+A2';
}
SignalGroups "First" {
    A3 = 'A1';// replaces global SignalGroup def
    A4 = 'A3+A2';// = A1+A2
    A3 = 'A2';// replaces A3 definition above
    A2 = 'A1';// replaces A2 Signal
    A5 = 'A2+A3'// = A1+A2
}
```

When both the unnamed SignalGroups (which is always used when present) and the SignalGroups named "First" are used together, the final set of SignalGroup definitions are resolved as:

A1 is the Signal A1

A2 is also the Signal A1

A3 is the Signal A2

A4 is a group that contains A1+A2

A5 is also a group that contains A1+A2

3.3.3 Applications of SignalGroups

SignalGroups can serve several purposes. By renaming individual signals, they provide a mechanism to reference the same signal by any number of alias names. This can also provide a mechanism to reduce the size of test data by using a short alias name in lieu of the longer signal names. Be aware that these two behaviors are somewhat at odds with each other — using aliases might provide a visual indication of the behavior of this signal at a particular point of the test operation, but using aliases as a short-hand signal renaming mechanism tends to make the test operation less easily readable (at least at the human level).

Grouping signals can also reduce test data volume by reducing the number of name references used in the Vectors. And it can make test operation more clear by indicating common assignment across a set of signals (such as a bus).

SignalGroups have another very specific purpose in the context of pattern reuse, and in hierarchical test creation using CTL concepts. In this context, test patterns created at a lower level of the design hierarchy, when integrated into a design, may find name conflicts between the core signals and the top level signals. This is shown in Figure 3.3, where the patterns generated for CoreA use the names "A" and "B", but when Core is integrated into DUT, these signals need to reference the Signals "B" and "C". By using the order dependent name resolution process, the patterns written for CoreA can be redirected to access the signal at the top level of the design.

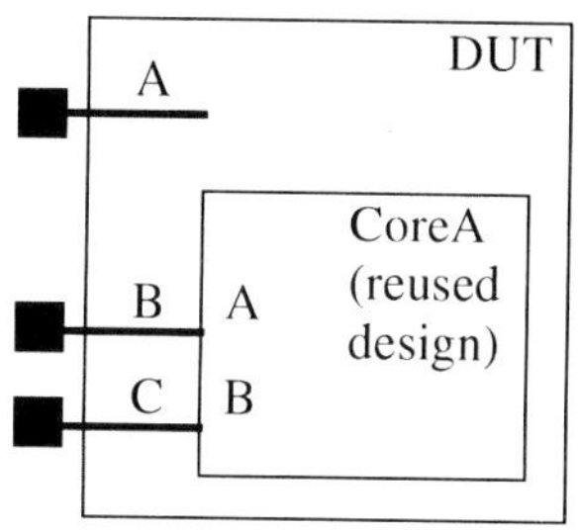

FIGURE 3.3 Name Conflict Resolution in Reused Patterns

The next section is generic to STIL and not specific to Signals constructs.

3.4 STIL Statement Constructs and Block Structure

The previous sections have presented two constructs of STIL — Signals and SignalGroups — without completely presenting how these constructs are organized in a STIL test program (although the examples have provided hints). Each of these constructs is organized in it's own block that starts with one of these STIL keywords — **`Signals`** or **`SignalGroups`**. The first token in a STIL statement is considered a *keyword*, and generally this leading keyword is reserved in a STIL environment for application in this context only.

STIL statements have two forms: a single statement form, and the block statement form. All statements in STIL support these two forms.

3.4.1 Single Statements Construct

The single statement form starts with a leading token, generally a STIL keyword, and may contain additional tokens depending on the statement. This single statement is terminated by a semicolon. For example:

```
(3.30)STIL 1.0;
```

3.4.2 Block Statements Construct

The block statement form replaces the semicolon with open and close curly braces, and allows the opportunity for additional statements to be defined inside, for example:

```
(3.31)STIL 1.0 {
(3.32)    Design 2003;
(3.33)}
```

The STIL Standard typically presents only one form for many STIL statements, either the block or the single statement form. For many statements, one of these forms may be more useful than the other. However, the intention of the definition is to allow any statement to be represented in either form.

3.4.3 Named Blocks

Several statements serve to define STIL blocks only; for example the **`Signals`** and **`SignalGroups`** statements. For many (but not all) of the Block statements, the next token before the opening curly brace is a user name, that represents a unique identifier for this block of information. For example, the SignalGroups block may contain a user name before the opening brace:

```
(3.34)SignalGroups local_A {... }
```

This name restricts the application of the SignalGroups defined in this block, as is presented in Section 3.3.2. The STIL Standard refers to this operation as “domain restrictions” on the application of this block, or “domain names” for these blocks.

If no user name is present on a block, the contents of that block are considered global in scope and the definitions present in that block are applied whenever the context of that block is appropriate. STIL allows only one global block of each type to be declared in a test program. The Signals block is always global; all Signals defined are always accessible (except when hidden by SignalGroup definitions in a named SignalGroups block that is referenced in the current context as presented in Chapter 6, Tying It All Together). There can only be one Signals block in any STIL test program and the Signals block is never named.

The next section discusses the motivation behind a language behavior and is informative only.

3.4.4 Reserving the Leading Keyword

STIL reserves the leading keyword on all statements defined in the standard. That means that those keywords are limited for use only in the context of that statement. What that *really* means is that keyword should not appear in other contexts parsed as part of a STIL program.

The motivation for defining STIL keywords as reserved was primarily to facilitate error detection and recovery for improper STIL files. By reserving keywords for application in specific contexts, improper use of the word is easily detected. Error recovery can be facilitated (particularly with missing braces) by an awareness of the expected context, or nearest expected context of the reserved word.

Reserving keywords also prevents reuse of the reserved name as a User name. This reduces the potential for generating obtuse STIL files by preventing these keywords from appearing at User name locations, confusing the parse operation if a syntax error is detected, and confusing the idle observer reading STIL code. Again, the STIL Convention that recommends double quoting all User names (see page 35) eliminates any conflict between User names and STIL keywords.

However, there are situations where keywords might appear in contexts outside of their expected use. One potential location is as part of the STIL Pattern data, covered in Chapter 5, Patterns. Due to the concatenation of WaveformCharacters into strings of characters, it is possible (although not at all easy) to create reserved words in this context. Therefore pattern data cannot be checked against reserved word usage, at least when parsing strings of WaveformCharacters.

3.5 Signals and SignalGroups Syntax

With this introduction to STIL syntax, the syntax of the Signals and SignalGroups sections can be presented. The following conventions will be used when STIL syntax constructs are presented. A word in **boldface** identifies a STIL keyword or a required syntax construct. A word in *italics* identifies a user-specified field. And when possible, an optional user-specified field will contain the phrase *optional_*.

So, the complete set of statements supported in the Signals block is:

```
(3.35)Signals {
(3.36)    signal_name ;
             – or –
(3.37)    signal_name {
(3.38)        Termination (TerminateHigh, TerminateLow,
TerminateOff, TerminateUnknown) ;
(3.39)        DefaultState (U, D, Z) ;
(3.40)        Base (Hex, Dec) wfclist ;
(3.41)        Alignment (MSB, LSB) ;
(3.42)        ScanIn optional_integer_value ; – or –
(3.43)        ScanOut optional_integer_value ;
(3.44)        DataBitCount integer ;
(3.45)}
```

The SignalGroups block has the same set of optional attribute statements, so without repeating those constructs, the SignalGroups block contains the statements:

```
(3.46)SignalGroups optional_signalgroups_block_name {
(3.47)    group_name = 'signal_reference_expression' ;
              – or –
(3.48)    group_name = 'signal_reference_expression' {
(3.49)        signal_and_group_attribute_statements ;
(3.50)    } }
```

3.6 Signal and SignalGroup Attributes

There are several attributes, or sub-characteristics, of Signals and SignalGroups, that define additional behaviors for these items. These characteristics affect either the behavior of these references during test execution, or the interpretation of these references when they are found in the test program. Therefore, these characteristics are collected into two groups: those that affect STIL interpretation, and those that affect the test execution only.

All attributes have a default value applied if the reference (Signal or SignalGroup) does not define a value for this attribute.

Some of these attributes are associated with the specific reference, in particular, the attributes that affect STIL interpretation. These attributes are applied only to that specific reference that defined the attribute; the attribute does not propagate down to other references or to the Signals themselves (unless defined on the Signals). However, the test-related attributes can only be defined once in an environment because these values *are* propagated down to the individual Signals in each reference where they are defined, and an individual Signal can have only one value of these attributes during the test operation. This is covered in detail in Section 3.7.2.

Table 3.3 identifies the attributes defined for STIL Signals and SignalGroups.

TABLE 3.3 Signal and SignalGroup Attributes

	Attribute	Values	Purpose
Test-related Attributes	Termination	TerminateHigh, TerminateLow, TerminateOff, TerminateUn-known	Defines a single value to terminate a Signal during the test operation.
	DefaultState	U, D, Z, ForceUp, Force-Down, ForceOff	Defines a single value to apply to an input Signal that is unused during the test operation.
STIL interpretation Attributes	Base	Hex or Dec, with WFClist	Provides default interpretation of WFClists assigned to this reference in the Pattern block.
	Alignment	MSB, LSB	Defines what bits to use for Hex or Decimal mapped groups.
	ScanIn	integer value	Identifies this reference is used as a scan input signal; the optional integer specifies the length of the scan chain associated with this reference.
	ScanOut	integer value	Identifies this reference is used as a scan out signal; the optional integer specifies the length of the scan chain associated with this reference.
	DataBitCount	integer value	Identifies a reference used in association with multibit waveforms, and identifies the number of bits associated with each complete data set in the multibit waveform.

The next section covers specific behavior of the test attributes, which is different than the other attributes but only necessary to understand if you are using these constructs.

3.7 The Test Attributes

Two of the attributes identified in Table 3.3, `Termination` and `Default-State`, do not affect the interpretation of the test program itself. The presence or absence of these attributes does not affect how the STIL data is interpreted. They do, however, affect the application of the test program during execution. For this reason, these attributes are identified as "test attributes". They also have slightly different semantics that differentiate them from the other attributes, as we will elaborate here.

3.7.1 Application of the Test Attributes

The purpose of the `Termination` attribute is to define a termination value to be present on the Signal during test operation. The intent of this termination value is to be passive, providing stability for this Signal during the test but not affecting the behavior of the test itself. Although, it is probable that the test cannot be executed properly if this termination value is missing in the execution environment. For instance, the ability (or the mechanism) to detect tristate values may be affected by the presence of a TerminateLow or TerminateHigh attribute, which indicates the effective presence of a pull-up or pull-down on this Signal during test.

The parameters of the Terminate attributes are specified as part of IEEE Std. 1450.2-2002, and are covered in Chapter 13, STIL Levels.

The purpose of the `DefaultState` attribute is to define a drive value to be applied to a Signal when that Signal is never referenced in a test. The DefaultState values are drive values only and are primarily intended to identify a state to hold unused inputs (or bidirectionals) during the test, but the language does not constrain this to inputs only if there is motivation to specify states for outputs (most likely, outputs would be held in a DriveOff value to prevent contention at test). When the DefaultState attribute is present, it is intended to provide a static, single value that this Signal will be held in during the entire test execution.

However, the STIL specification defines one additional application of the DefaultState attribute, if present. This attribute is applied to unused Signals in a Procedure call, as identified in Chapter 11, Procedures and Macros. This application of the DefaultState is largely a dynamic operation (the set of unused Signals can vary across different Procedures), which makes static application of this value only possible for Signals that are truly unused for the entire test set.

Also, be aware that the STIL Design Extensions further expand on the use of the DefaultState, to support additional dynamic application of these values as part of the test execution.

STIL Conventions

Because of the dynamic and static behavior of the `DefaultState` attribute, use this attribute with care. For instance, if the intent of this attribute is to present a static state on a Signal during test execution, make sure that this Signal is used uniformly in a test environment. In particular, make sure that if this Signal is used in a Pattern then it is also consistently used in all Procedures called by this Pattern. This is to avoid an unintentional application of the dynamic DefaultState construct in a Procedure that did not use this Signal (if the Procedure expects the DefaultState to be applied for proper Procedure operation).

3.7.2 Propagation of the Test Attributes

The test attributes are defined in the STIL specification to be present only once per Signal. Furthermore, these attributes only have a meaning when associated with a Signal. However, these attributes may appear on SignalGroup definitions. This implies that these attributes must propagate down through the SignalGroup members and eventually are found on the Signals themselves.

```
Signals {
    "A1" In { Termination TerminateHigh; }
}                                              Error on signal A1
SignalGroups {
    "A2" = '"A1"' { Termination TerminateLow ; }
}
```

FIGURE 3.4 Incorrect Terminate Environment

This behavior is unique to the test attributes. It is necessitated by the issue that these attributes only have meaning when associated with the Signals. All other attributes affect only the reference where they are defined. The attributes do not carry down to the individual Signals, nor do they propagate upward to be contained in another reference.

Because the test attributes propagate down to Signals, there are two recommended strategies for defining these attributes. These usage models are presented next. Figure 3.4 is an example of an improper application of the Terminate statement.

3.7.3 Two Usage Models for the Test Attributes

To minimize the potential for conflict between multiple values of the test attributes associated with an individual Signal, there are two recommended styles of defining these attributes. These two styles of defining the test attributes fall roughly into two camps of test operation: part-of-the-test-fixture, and part-of-test-execution. Depending on your perspective of how this issue is supported at test, your strategy to apply these attributes may follow one of these models.

The part-of-the-test-fixture camp supports only one value of this attribute, which may even be physically implemented in the test fixture itself. The value of this attribute cannot change during the test execution. This camp would define the attribute on the Signal itself, in the Signals block, and would not define it in any other location. This identifies one location to specify the attribute, facilitating both the definition and the subsequent identification/localization of the value.

There is one twist to this option: the attribute may be defined in the global SignalGroups block instead of the Signals block, since there can be only one of these blocks. While this option is allowed (the STIL specification cannot prevent it), this strategy would be recommended only in an automated flow (a flow that generates the global SignalGroups) to minimize risk that the attribute might be found in two locations.

The part-of-test-execution camp supports some amount of dynamic behavior on these values; they are static during a particular test execution, but the value may change during execution of different test sequences. This camp would define the attribute only in named SignalGroup blocks, where the named SignalGroup blocks are referenced as part of each test execution (this is presented in Chapter 6, Tying It All Together). This allows the value of the attribute to change between PatternBursts. But there is risk here: it is easy to define the attribute in two named SignalGroups blocks and incorrectly reference both SignalGroups blocks in one PatternBurst, which would require corrections before the STIL data could operate.

There are several STIL behaviors in addition to these test attributes, that fall into similar usage models: behaviors to be defined only once for the test, and behaviors that can be changed at some level. The strategy for supporting these dichotomies is generally similar to what has been presented here: for single behaviors, use the "global unnamed" block to define these behaviors in a single location, for dynamic behaviors use named blocks and the referencing mechanisms to identify which one is active during the test.

3.8 WGL Signals

At the end of some chapters, we will present very briefly the constructs in the Waveform Generation Language (WGL) that support the behaviors of the STIL constructs in the chapter and compare and contrast the two environments. We start this discussion with Signals in WGL.

The WGL environment defines a block to contain the signals of the test, and places that block up front in the order of test information — very similarly to STIL. The WGL block even has the same name — *signals*.

WGL signals have similar definition of types (*input*, *output*, and *bidir*) and flexibility for declaring bussed signals (using the same square brackets and two-period ellipsis notation), but more rigorous constraints when handling a collection of signals declared as a bus.

WGL supports an additional construct called *mux* that identifies when multiple signals are meant to be tied together at test. STIL supports WGL mux behaviors with enhanced waveform representation. There are positive and negative aspects of each style of behavior. In STIL the complexity of the waveform (or collection of waveforms) may identify special considerations at test, but the identification of the support necessary in the test environment to address this complexity may not be straightforward. Conversely, the WGL *mux* construct requires interpreting two independent streams of test data (columns in the WGL patterns, when we get there) in a dependent fashion to establish the actual signal behavior on a per-cycle basis.

WGL also supports collecting groups of signals and referencing the collection with a group name. But WGL groups are declared after the signals as part of the same WGL *signals* block. There is no separate block for groups, and there is no notion of named collections of groups that may be applied to different contexts. This complicates pattern reuse in a WGL environment, but conversely prevents confusion in the situation when the same group name is declared in multiple STIL SignalGroups blocks and which is active in a pattern is difficult to identify. WGL group constructs support only an additive assembly of groups from other signals or groups. There is no notion of removing a reference from a previously-defined collection.

WGL defines a series of pre-declared groups *ALL*, *ALLINPUT*, *ALLOUTPUT*, *ALLBIDIR*, and *ALLMUX*. These definitions do not exist in STIL by default and need to be included in the SignalGroups definitions if desired.

CHAPTER 4 Timing

Standard
Test Interface
Language

Automated Test Equipment relies on regular, repeated sequences of events. Likewise, Automated Test Pattern Generators from CAD tool environments expect a generally limited sequence of repeated events to execute test programs. These regular event sequences are called Waveforms in STIL, and are contained under the Timing block of a STIL program.

4.1 The Timing Block

STIL Timing information is organized under a **`Timing`** block. There may be multiple Timing blocks in a STIL test program following the restrictions defined above.

The Timing block contains definitions of event behavior across all Signals currently active in a test operation. These definitions are collected in WaveformTable blocks.

A fundamental restriction of the STIL environment is that all activity across all active Signals be defined in a single WaveformTable; as we will present in the next Chapter only one WaveformTable is active at a time. However, each

Signal may contain multiple behaviors (Waveforms) in one WaveformTable. The decision of how many behaviors to define on each Signal in each WaveformTable, and how many WaveformTables to define in one Timing block, is often dependent on the application environment and is discussed in detail in Chapter 9, Partitioning Timing.

All WaveformTables used in a set of Patterns must be defined in one Timing block as well. This is covered in detail in Chapter 6, Tying It All Together. While all WaveformTables do need to be defined in one Timing block, STIL supports several mechanisms to incorporate all the WaveformTables (from potentially multiple blocks) into one block. These mechanisms are presented in Chapter 10, Advanced Timing. The Include mechanism (Chapter 7, Files and the Include Statement) can also be used. Until the more complex features of the language are presented, we will limit the Timing to be contained in a single block.

There are many parts to a complete WaveformTable. In the following sections, we will assemble the WaveformTable from the bottom up, starting with the lowest level: Events.

4.2 Event Statements

Waveforms consist of a sequence of events, for example the description of a high-going pulse needs at least the definition of a rising edge followed by the definition of a falling edge. Each edge or any change of behavior in a waveform is represented by an event. An event consists of two pieces of information: a reference to a time, and a **state**. When that time occurs, the state is asserted.

An event statement contains two fields: a time value and a **state**. These two constructs are combined in a single statement as shown below:

```
(4.1) 'time_expression' state_value;
```

The 23 state values available to a STIL test are listed in Table 4.1. Each state has very specific behavior for each value and purpose. STIL provides both a single character keyword and a full name for each state; either may be used to specify a state in an event statement.

TABLE 4.1 States in STIL

Type	State	Behavior
Drive	`D` *or* `ForceDown`	Assert a "low" state on the Signal
	`U` *or* `ForceUp`	Assert a "high" state on the Signal
	`Z` *or* `ForceOff`	Remove or disconnect a drive state
	`P` *or* `ForcePrior`	Reapply the last D or U asserted
	`N` *or* `ForceUnknown`	Provide a drive state, but unknown value
Compare	`L` *or* `CompareLow`	Detect a "low" state on the Signal
	`H` *or* `CompareHigh`	Detect a "high" state on the Signal
	`X, x` *or* `CompareUnknown`	Remove or discontinue a detect state
	`T` *or* `CompareOff`	Detect a mid-band state on a Signal
	`V` *or* `CompareValid`	Detect either a "high" or "low" state
	`l` *or* `CompareLowWindow`	Detect a "low" region on the Signal
	`h` *or* `CompareHighWindow`	Detect a "high" region on the Signal
	`t` *or* `CompareOffWindow`	Detect a mid-band region
	`v` *or* `CompareValidWin-dow`	Detect either a "high" or "low" region on the Signal
Expect	`R` *or* `ExpectLow`	Expect a "low" state
	`G` *or* `ExpectHigh`	Expect a "high" state
	`Q` *or* `ExpectOff`	Expect a "mid-band" state
	`M` *or* `Marker`	Provides a timing reference point
Indetermi-nate	`A` *or* `LogicLow`	Unknown direction but a "low" logic state
	`B` *or* `LogicHigh`	Unknown direction but a "high" logic state
	`F` *or* `LogicZ`	Unknown direction but a "mid-band" logic state
	`?` *or* `Unknown`	Nothing is known about the state

The STIL standard confuses the term *event*; it uses the term *event* to refer to both the time-and-state statement, and to the stand alone state.

4.2.1 States and Levels

The STIL states refer to a "high", "low", or "mid-band" value in order to perform measurements or to apply values, without actually defining what these values are. Actual values are typically technology-dependent and may vary across the Signals in a single design.

The specification of the voltage/current values is provided by the IEEE Std. 1450.2-2002 extension and is covered in Chapter 13, STIL Levels (Section 13.3.2, on page 228).

4.2.2 ForceUnknown, 'N'

The ForceUnknown may be considered or treated as a "drive X" from a simulation perspective. It is considered to be a conservative state, especially when simulated as an X. An X in simulation is expected to behave in a conservative fashion, pessimistically representing the effects of either a '1' or a '0'. This means, when present at the test program, a ForceUnknown may be replaced by a ForceHigh or a ForceLow without affecting the test operation. Note in the STIL specification, ForceUnknown is defined as an unresolved event (which it is), but in Table 4.1 it is placed with the drive events. From the test perspective it is a drive event with a "don't care" level so it is placed with the drive events in this table.

STIL Conventions

The "N" state may be resolved during test operation as any permissible drive level, since it explicitly identifies a "don't care" situation. Ultimately the decision of how to resolve the "N" state may take into consideration the characteristics of the specific pin, the ATE electronics, or even the previous state of the Signal. In lieu of a motivation for a drive value to use in place of the "N", choose a consistent value such as "0".

4.2.3 The Difference Between Compare and Expect States

The STIL compare states identify points during the test execution where the test equipment performs a measurement. For reproducible tests, test measure points are specified at locations where the output value being measured is assured to be at that state. The position or timing of these measurements are typically guardbanded by some amount to assure reliable response. This guardband is factored for both behavior across the range of acceptable device operation, and the acceptable variation of test equipment operation.

The *intent* of Expect states is to reflect the actual device timing behavior. Expect states are not necessarily a part of the test data, but may be present to validate device operation in other contexts (for example, against simulation responses) or provide information about the actual timing of the device, perhaps with only a single time context (e.g, best-case or worst-case timing), or perhaps across a range of conditions. How Expect data is applied, or if it is applied at all, is dependent on the application receiving this data. It may be ignored entirely by the application. It may be used to evaluate or check the value of the guardband region. Figure 4.1 presents a representation of a device response with both Expect and Compare information.

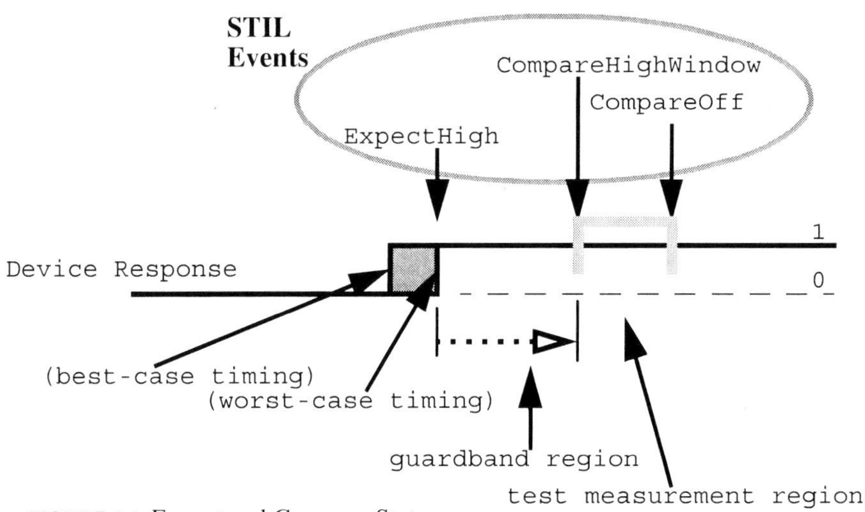

FIGURE 4.1 Expect and Compare States

4.2.4 The Relationship Between Drive and Compare States

Drive events and compare events define separate sets of operations, or distinct "tracks" of behavior. Always remember that these two tracks are completely independent. For example, it is possible to assert a 'D' event to drive a Signal low, and immediately follow this by an 'H' event to test that this same Signal is high. Of course, unless you are manipulating the test levels (described in Chapter 13, STIL Levels), this operation should result in a failing test. And also remember: Z events are drive operations, X events are compare operations, and setting one of these events will not affect the current state of the other track.

4.3 Basic Time Expressions

The time value in an event statement represents the offset from the start of the current Waveform at which the specified state is asserted. These time values may be represented by a constant value or a complex expression that may be dependent on both the values of variables and of the timing of other events. Relative time events and event expressions using variables are described in Chapter 10, Advanced Timing. We will start our discussion of time values with simple, constant expressions.

Constant time expressions consist of a number followed most commonly by a scale factor, and a lowercase 's' to indicate the SI unit of **seconds**. Scale factors are: E, P, T, G, M, k, m, u, n, p, f, and a. The number may be an integer value or a floating-point value containing a decimal point. The number may be negative as well.

Time expressions are always enclosed in single quotes when used inside waveforms, for example: '10ns'.

STIL Conventions

Do not put a blank (or any whitespace) between the number and the scale factor of a time expression. A scaled time value (a number with a scale factor and an 's' for seconds) is considered a single token in the STIL environment and therefore cannot have whitespace in it.

4.4 *Waveforms*

The STIL definition of a waveform is a stream of events. Waveforms are constructed in STIL syntax by:

```
(4.2) Signal_references { WFC_references { events; }}
```

There may be zero or more event statements in a single Waveform definition. Waveforms are referenced by WaveformCharacters. WaveformCharacters are defined in the next section.

The signal_references are resolved specifically in the Timing block. This is covered in Section 4.10.1.

Waveforms define an infinite variety of behaviors, but can be generally categorized in several different ways. These types are not part of the STIL definition, but are useful when discussing STIL behaviors. These categories are presented in the next section.

STIL Conventions

A waveform containing no events would not appear to be very useful. However, because of the STIL behavior of event persistence, sometimes an empty waveform is necessary to change the currently referenced waveform and remove some set of events in the previous behavior but maintain other events in the previous waveform still active. This behavior is covered in more detail in Section 4.7.

The next two subsections present some waveform characteristics and test terminology, not restricted to the STIL environment but background for the general application context.

4.4.1 Types of Waveforms

Following along the lines of the types of Signals presented in Section 3.1.3, waveforms can be categorized into similar types. These types are presented in Table 4.2. Often, but by no means necessarily, there is strong correlation between the Signal type and the Waveform types associated with that Signal.

TABLE 4.2 Waveform Types in STIL

Name	Behavior
`In`	Waveform contains only drive events.
`Out`	Waveform contains only compare events.
`InOut / Mixed`	Waveform contains a mixture of event types, both drive events and compare events. Certain drive events may be ignored when classifying In or Out waveforms even when they contain a mixture of events. See "Special Events" on page 78.
`Indeterminate`	Waveform contain events that do not define a definite direction.

Classifying waveforms by type is dependent on the requirements of the application, depending on the constraints and behavior of the environment. Mixed or InOut waveforms may be classified as "pure" In or Out waveforms by some applications depending on the sequencing or specific type of each event and the requirements of the application. For example, a drive event preceding a compare event to the same state may be considered a "pure" In waveform because the compare event is only validating the drive operation. And tristate or mid-band events (which are defined to be drive events but serve the function of removing drive states) associated with compare events later in the waveform may still be defined as Out waveforms as the intent of the waveform is to determine device response without ATE effects. These decisions are very much application specific. It is important to realize that there are no specific restrictions on this classification process.

4.4.2 Waveform Styles

The number and types of events define some conventional styles of waveforms as presented in Table 4.3. These terms are generally recognized across the ATE industry and are not restricted to STIL applications but are defined here in STIL terms.

TABLE 4.3 Representative Waveform Styles

Name	Behavior
`NRZ`	Waveform contains one drive event — "Non-Return-to-Zero", which is a reference to the non-pulsed behavior of this waveform.
`NRZ-zero`	A sub-classification of NRZ waveforms. Waveform contains one drive event at time-zero in the cycle. This waveform may be specifically identified for environments where NRZ-zero waveforms have different characteristics from delayed NRZ waveforms (next).
`DNRZ`	Waveform contains one drive event not at time-zero (Delayed NRZ).
`RZ / RO / pulsed`	Waveform contains at least two events that define a pulse. There may be an event at time-zero. The acronyms refer to "Return-to-Zero" and "Return-to-One", which is a reference to the pulse type (actually the end state) of the waveforms.
`Multiple Pulsed`	Waveform contains multiple drive events creating multiple pulses.
`Strobe`	Waveform contains one or more compare events.

STIL Conventions

Most events in a STIL Waveform occur inside the boundaries of the Period timing. However STIL supports events that happen after the end of the current Period (into the next Period), and events that happen before the start of the current Period as well. While STIL allows these constructs, considerations must be made about whether the application environment will also support these constructs.

Consider carefully the specification of negative times in waveforms. Negative times may be explicitly defined, to occur before the start of the Period. They may also occur in more subtle contexts, for example when an event is specified to occur before another event. Many environments may not support negative values or values that are specified to occur before the start of the current Period. Also beware of a negative value in the very first Vector (where there is no previous time period to back into), which may be expressed as occurring at the start of that Vector, effectively ignoring the negative value.

Keep associated events together in a waveform. When events occur across cycle boundaries, sometimes it may not be apparent what events should be associated together or maintained in a single waveform description. Some ATE architectures require related events to be defined as a unit. Also, ATE resources tend to be allocated on a Waveform basis, and by maintaining these events in a single Waveform a common timing base is generally applied for all events. When the events occur in separate Waveforms, particularly from different WaveformTables, the time base may not be in common across all these events. This in turn may result in increased variation in timing of the events on one Signal. See Section 4.8 for additional elaboration on this issue.

When events in one pattern cross over to another pattern boundary, be aware of test restrictions that may be present between the two sets of events. For example, many test environments do not allow an event from a previous pattern to extend beyond the first event, or the first event of that type, in the next pattern.

4.5 WaveformCharacters

A WaveformCharacter is a reference to a predefined set of activity (a Waveform) on a Signal. The actual behavior of that Signal is dependent on both the current active WaveformTable and the WaveformCharacter definition for that

Signal in that WaveformTable. A WaveformCharacter is a single alphanumeric character (in some circumstances presented later, it may take more than one character to resolve a Waveform's activity completely).

WaveformCharacters are used in the Patterns, and form the bulk of the test data. They identify the activity for each Signal during the test. The application of WaveformCharacters to Signals is covered in Chapter 5, Patterns.

The activity defined on a WaveformCharacter is defined by the Waveform definitions. There is no pre-established behavior for any WaveformCharacter, and it is possible to use the same WaveformCharacter in different WaveformTables to generate different Waveforms, or across different Signals in the same WaveformTable. You should not assume that you understand the behavior of a Signal from the WaveformCharacter being used, but should always validate that the Waveform defined for that Signal with that WaveformCharacter in that WaveformTable is the behavior you anticipate.

While you should not assume a behavior from a WaveformCharacter, there are conventions of WaveformCharacter use. These conventions are listed in Table 9.1, "Conventional WaveformCharacter Representations," on page 150.

4.6 Merging Common Waveforms

One STIL Waveform definition is not constrained to a single sequence of events. Multiple Waveforms may be combined into a single Waveform definition.

The motivation for merging Waveforms is as a representational construct only. There is no functional difference, and there is no operational difference when Waveforms are merged to a single definition. This option may save a little space, and it makes the WaveformTable a bit more compact which can make it more readable. But there is absolutely no functional difference between a merged Waveform and an equivalent set of events in separate Waveforms.

Another motivation for merging Waveforms may be to represent a collection of Waveforms as a single entity, and as such identify the collection as a single unit of test resources. This may facilitate mapping of this information into that tester environment, or may facilitate incidental operations, like the generation of messages that reference this timing information, which identifies these quantities with a direct relationship to the tester resources.

To combine two or more Waveforms into one definition, several constructs must be present. First, multiple WaveformCharacters must be defined as a list of WaveformCharacters in the Waveform statement. Then for each event that has a different behavior across the WaveformCharacter, a unique mapping of states to each WaveformCharacter is defined. Multiple state values are presented when the Waveforms being merged do not have the same behavior at this time; if the event is common across all the Waveforms then a single event definition is sufficient. Multiple states are specified syntactically by separating the states with a slash character ('/'). The addition to the event statement can be diagrammed as:

```
(4.3) 'time_expression' state_value / state_value /...;
```

All state values in this event statement transition at the same time specified by the single time_expression. Only one state is in effect at a time because of the relationship of the WaveformCharacters to this list of states.

One typical scenario of a merged waveform is the representation of a single edge (NRZ) waveform where both the drive-low and drive-high waveforms are combined. The merged waveform definition requires two WaveformCharacters - one for the drive-low, and one for the drive-high. The single event in this waveform (remember this is an NRZ example) presents the first state for the drive-low WaveformCharacter, and the second state for the drive-high WaveformCharacter. The list of states in the event are ordered to match the sequence of WaveformCharacters presented; the first state is associated with the first WaveformCharacter, etc. This is shown in the next example:

```
(4.4) 01 { '0ns' D/U; }
```

The advantage of this representation is the emphasis on common timing between these two sets of waveforms. In some ATE environments this form of definition may reflect the ability of the ATE to use common timing resources for both waveforms. But this is not a requirement of this representational form — this is only a representational option in STIL.

Over-using the '/' operator can result in a situation called *Waveform over-merging*. This is discussed in "Waveform Grouping/Merging" on page 151, Section 9.4.

STIL Conventions

When the '/' operator is present in an event statement, the number of states presented in that statement must match one-for-one with the number of WaveformCharacters defined for that waveform. When only one state is present in the event statement, all waveforms defined (one for each of the WaveformCharacters) will apply that event.

4.7 Event Ordering and Persistence

Waveform Events are classified into four categories as identified in Table 4.1. The events of each type in a Waveform must be presented in increasing time sequence in the Waveform, but events across different categories can be mixed in any order. Hopefully, the next example will make this more clear:

```
(4.5) P { '10ns' U; '20ns' D; }// correct - drive events
(4.6) N { '20ns' U; '10ns' D; }// incorrect - wrong order
(4.7)    // The following example is also correct
(4.8)    // because L is a compare event and the other
(4.9)    // two events are drive events.
(4.10)P { '30ns' L; '10ns' U; '20ns' D; }
```

There are no specific requirements on what events, types of events, or combinations of events may be defined in any particular Waveform. Waveforms may consist of events from only one of the categories or may contain a mixture of events from several different categories.

Waveform behavior persists beyond the end of the current cycle. As we will present in the next Chapter, if the WaveformCharacter on a Signal is not changed in the next Vector then the behavior of the current referenced Waveform will repeat.

But changing the WaveformCharacter in the next Vector (and consequently the referenced Waveform) does not necessarily change all the behaviors on the Signal. In particular, the *last event* of each type in Table 4.1 will persist in the next Waveform until it is specifically changed by another event of that type in that (or some subsequent) Waveform.

This behavior may result in some unexpected behavior in an overall sequence of Waveforms if the persistent operation of the last event is not considered. It also places requirements or additional expectations on the definition of Waveforms in order to properly define Waveform operation and sequences of Waveforms.

The last event of each type continues to be asserted from one Waveform and into the next effective Waveform, until changed by the first event of that type in a subsequent Waveform. The impact of this behavior can be very subtle. It is important to realize that the current WaveformCharacter does not tell the

```
Waveforms {
    output_Signal {HL  { '70ns' h/l; '90ns' X; }
                   Z   { '0ns' Z; }
    }
}
```

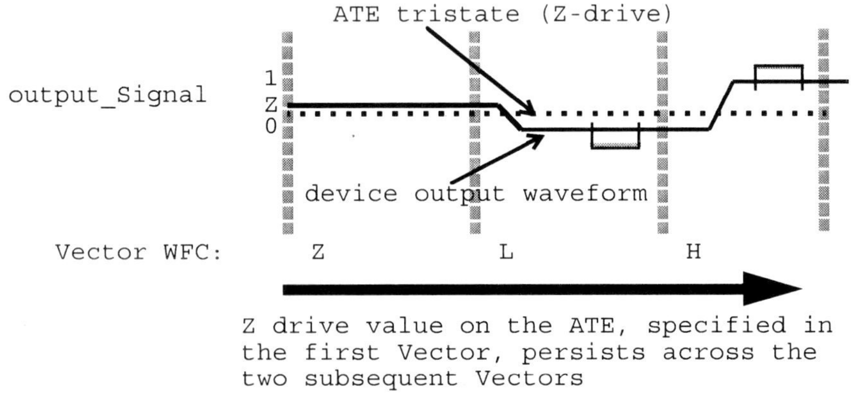

FIGURE 4.2 Persistence of a drive state in subsequent Vectors

entire story of the state on a Signal; prior history can impact the current Waveform as well.

Bidirectional signals, which support both drive events when the signal is being used as an input, and output events when the signal is being used as an output, may depend on this persistence behavior. Consider an environment that uses waveforms in sequence to perform separate test operations: the first waveform establishes input states across all the signals being driven, and the subsequent waveforms measure the output states on all signals being compared. Bidirectionals may at some times be driven in this context, and at other times be compared. Therefore the waveforms used for these signals may need to expect the previous waveform to establish requirements for the next operation. This is shown in Figure 4.2. The grey vertical lines represent the boundaries between waveforms.

The next section presents an example of the STIL language application that affects test operation and interpretation.

4.8 Maintaining Related Events

STIL supports much flexibility for representing timing information. The following two examples present scenarios whereby associating events occurring in the "next" Vector in a Waveform defined in the previous Vector results in a cleaner representation of the information, potentially reduces difficulties in test transportation, and maintains a better perspective on the intent of the operations occurring.

4.8.1 End-Strobe Relationships

Figure 4.3 presents an output window strobe that straddles cycle boundaries.

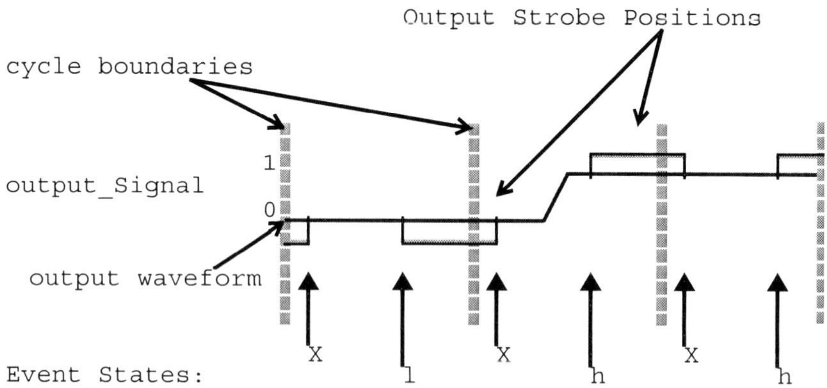

FIGURE 4.3 Window Strobe Across Vector Boundaries

Figure 4.4 presents one STIL representation of window strobes across Vector boundaries.

```
STIL 1.0;
Signals { output_Signal Out; }
Timing { WaveformTable one {
    Period '100ns';
    Waveforms {
        output_Signal { HL { '10ns' X; '70ns' h/l; }}
    }
}} // end Timing
```

FIGURE 4.4 Representing the Strobe Events In Time-Sequence

So why is this a problem? If your ATE environment supports this definition, then there is nothing wrong with this. But as stated previously, the sole purpose of the 'X' event is to terminate an active strobe measurement. It serves that purpose here, but it terminates the strobe that started *in a previous cycle*.

The interesting situation here is the middle cycle in Figure 4.4. The 'X' event terminates an output check for an output low at the start of the cycle, and starts a check for an output high condition at the end of the cycle. Two different values in one cycle might require special support in the ATE environment, at least if the ATE differentiates which part of the test cycle actually failed and how. In general, because the 'X' event is terminating an operation that started in a previous cycle, there's no direct way to associate the end-strobe activity with any definitive start-strobe operation.

Figure 4.5 presents the other way of representing this information.

```
STIL 1.0;
Signals { output_Signal Out; }
Timing { WaveformTable one {
    Period '100ns';
    Waveforms {
        output_Signal { HL { '70ns' h/l; '110ns' X; }}
    }
}} // end Timing
```

FIGURE 4.5 Representing the Strobe Events In Associated Order

In this representation, the 'X' event does terminate the start-strobe event. The issue here is that the end-strobe event is defined after the end of the test cycle. It still requires ATE support for this construct as well. But the relationship of the end-strobe event is clearly associated here with the start-strobe event.

4.8.2 Delayed Pulse or NRZ Waveforms

Figure 4.6 presents an input event sequence on a Signal.

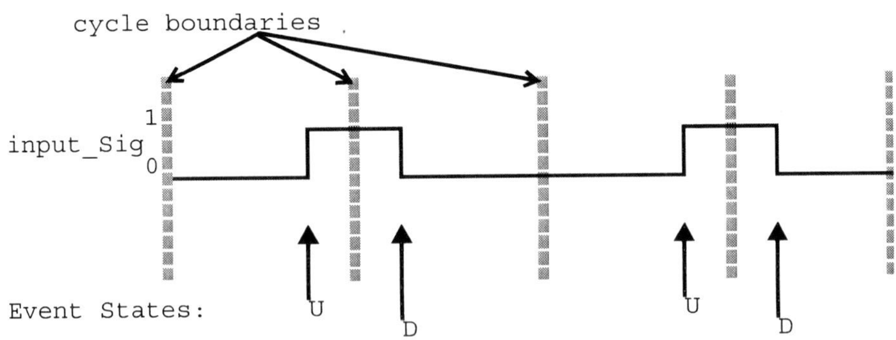

FIGURE 4.6 Input Events

Figure 4.7 presents one STIL representation of the input events in Figure 4.6. In this representation, the combination of the rising edge and the falling edge behaviors are allocated for each different time that *one* of these events occur. That means that even though the waveform in Figure 4.6 only contains one rising edge time and one falling edge time respectively, because of the pairing of events (an assumption of the ATE environment requirements), several additional Waveforms and WaveformCharacters are consumed.

```
STIL 1.0;
Signals { input_Sig In; }
Timing { WaveformTable one {
    Period '100ns';
    Waveforms {
        input_Sig { AB{ '80ns' U/D; }}
        input_Sig { CD{ '20ns' U/D; }}
    }
}} // end Timing
```

FIGURE 4.7 Representing the Input Events

Again, this representation is not a problem if your ATE environment supports this number of waveform definitions. However, it is not uncommon in ATE environments that each different waveform time sequence requires a separate WaveformTable. Rather than being able to present multiple sets of WaveformCharacters in a single WaveformTable, an alternate definition might require these two sets of Waveforms to be placed in two separate WaveformTable definitions. This causes the Pattern information (which is presented in detail in Chapter 5, Patterns) to switch between these two WaveformTables for each Vector — a situation known as "WaveformTable thrashing".

Figure 4.8 presents another way of representing this input behavior.

```
STIL 1.0;
Signals { input_Sig In; }
Timing { WaveformTable one {
    Period '100ns';
    Waveforms {
        input_Sig { AB{ '80ns' U/D; '120ns' D; }}
    }
}} // end Timing
```

FIGURE 4.8 Representing the Input Events As a Pulse

By representing this behavior as an offset pulse, where the trailing edge of the pulse occurs in the next test cycle but is defined as part of the current cycle activities, several things happen:

- Waveform definitions are reduced in this WaveformTable definition, which correlates to reducing tester resource requirements.
- The timing of the falling edge is correlated with the rising edge by virtue of a common Waveform definition. The characteristics of the pulse are much more apparent.
- The "WaveformTable thrashing" situation, discussed in the previous scenario, is eliminated here because only one WaveformTable needs to be defined (at least for this Signal). The Vector data in the Pattern section alternates between A and B WaveformCharacters, but both of these behaviors are defined as presented in Figure 4.8.

STIL Conventions

There are other motivations for presenting the data in the more cohesive formats. For example, test environments can take advantage of these formats if this information is manipulated during test. If the timing of a pulse is incrementally modified during a period of repeated (looping) tests, to observe device sensitivities to the pulse timing, these timing changes are facilitated when the timing is expressed in this format and not as a series of independent (or apparently independent) waveforms.

4.9 Special Events

There are several types of special events that may be necessary in certain circumstances, or indicate particular behaviors to support execution of the test in a test environment. These events are presented here.

4.9.1 Time-Zero Event

Some test environments require a state to be specified at the start of each waveform, especially for In-type waveforms as described in Table 4.2. For single-event (NRZ) waveforms, often the single event occurs at the start of each waveform, or "time-zero" in this waveform.

Delayed NRZ waveforms, and pulsed waveforms (where the pulse does not start at the beginning of the waveform) may still need an explicit event to be specified at the start of the waveform timing. For this purpose, STIL defines the P state for In-type waveforms. The P state is *intended* to be used in delayed In-type waveforms to specify the initial event if that specification is necessary in the STIL application.

Remember that the P state also implies that the ATE driver will be enabled (to the previous active-drive state); be careful with the P state if a tristate value needs to persist into a Vector that uses this Waveform.

4.9.2 Tristate Events

Bidirectional signals require explicit manipulation of the ATE drive value in order to avoid test contention between the ATE driver and the device driver. The STIL Z state is used to identify when the ATE driver must be disabled; an active drive value (U or D) or the P state is used to identify when the ATE driver is to be enabled again.

In some ATE environments, enabling and disabling the ATE driver occurs in a different timing context (or with different limitations) than other drive events. STIL does not make these differentiations explicit; the Z state is manipulated just like any other STIL drive state.

STIL does not make any assumptions about waveform behavior based in the style of the waveform. Out-style waveforms (again, as defined in Table 4.2) will not implicitly disable the ATE driver; the waveform must contain an explicit Z state (or the Z state may have been applied in a previous Vector).

If a bidirectional signal, in an Out-style waveform, requires a Z state to disable the driver as part of the Waveform description, technically the waveform type becomes an Inout style. This issue impedes the direct classification of waveforms, as bidirectional waveforms with output strobes will always have this Input event also present. Therefore some application contexts will ignore the Z drive state on a bidirectional otherwise-Out waveform to permit classification of a strobed output waveform as an Out-style waveform.

STIL Conventions

Do not forget about the Z state on bidirectionals! Some ATE environments may provide an implicit tristate operation, particularly when Signals are being strobed, but it is not safe to rely on these implicit operations as part of a transportable STIL file. An ATE environment without the implicit behavior requires an explicit representation of that behavior, and it is never incorrect to be explicit with the Z.

4.10 The Rest of the Timing Block Syntax

The basic syntax of the Timing block, down to the Waveforms assigned to Signals, looks like:

```
(4.11)Timing optional_timing_name {
(4.12)   SignalGroups signalgroup_block_name ;
(4.13)   WaveformTable required_wft_name {
(4.14)      Period time_expression;
(4.15)      Waveforms {
(4.16)         Signal_references { WFC_references {
(4.17)            events; }}
(4.18)         }
(4.19)      }
(4.20)   }
(4.21)}
```

There may be multiple Timing blocks in one STIL test program as long as each block has a unique name or one of the blocks has no name at all. Each Timing block may contain multiple WaveformTable declarations; each WaveformTable must have a name and these names must be unique across all WaveformTables contained in one Timing block. When named Timing blocks are present, the timing information they contain is applied only when it is referenced from a PatternExec. This is presented in Chapter 6, Tying It All Together.

Each WaveformTable contains at most one Period statement, and at most one Waveforms block. As we shall see later (Chapter 10, Advanced Timing), some circumstances allow the absence of these statements but require the presence of additional constructs that will be presented there.

All waveforms, presented in Section 4.4, are collected under the one Waveforms block in a WaveformTable.

4.10.1 Period Statement

Waveforms across Signals are collected under a WaveformTable (inside a Waveforms block), as shown in the previous syntax. In addition to the individual Waveforms, a single Period statement is required in each complete WaveformTable.

The Period statement contains a single time value that is the amount of time to advance for this collection of events. This statement identifies the cycle time of a Pattern Vector (discussed in the next chapter) when this WaveformTable is applied. Individual events in particular Waveforms may exceed this time value (in which case the event is executed into the next cycle), but most commonly the last event in the Waveforms is finished before this period time has elapsed. Most importantly, the next cycle does not start until the period of the current cycle has elapsed.

4.10.2 Signal Reference Resolution in Timing

Each Waveform contains a reference to a set of signals. This was identified with statement (4.2), which is repeated here with an emphasis on the signal references:

```
(4.22) Signal_references { WFC_references { events; }}
```

The signal references made in this statement may take all the forms of references supported by STIL: a direct reference to Signal name or array, a reference to a signal group, or a signal expression itself, enclosed in single quotes (a "group-on-the-fly").

These references are resolved in a special way, specific to each Timing block and independent of other resolution contexts in STIL. If a SignalGroups block that has a signalgroup_block_name (a "named" block) contains signal group names that are necessary to resolve these signal reference names, then that SignalGroup block name must be explicitly identified with a SignalGroup reference statement in this Timing block. The context of this statement was shown at statement (4.12), and the statement is repeated here:

```
(4.23)    SignalGroups signalgroup_block_name ;
```

Each SignalGroups block with a signalgroup_block_name is specified with a separate statement. This statement may appear multiple times to identify all named SignalGroups blocks that are required to resolve the signal_references in this Timing block.

The set of signal groups defined across all the named SignalGroups blocks creates a special collection of signal groups names. Any of these group names may be used in a signal_reference in statement (4.22). Each signal group name in this collection must be unique across all of the *named* SignalGroups that are referenced by statement (4.23). STIL does not allow a signal group name to be declared in two different named SignalGroups blocks, *and* to reference both of those groups from one context, even if the signal group contents are identical.

In addition, the Signal names, and the names of signal groups defined in the unnamed SignalGroup block, can be referenced in the Timing without any special reference statement. These names are always accessible.

However, the Signal names and signal group names in the unnamed SignalGroups block may be replaced by names found in one of the named SignalGroups blocks that was referenced with statement (4.23). The signal group names defined in named SignalGroup blocks have the capacity to replace the names that are always accessible, but they cannot replace names found in other named SignalGroups blocks.

This allows each Timing block to resolve signal_references independently. This model of signal name resolution is used by several other STIL contexts as well; whenever a SignalGroup "reference statement" such as statement (4.23) is present in a STIL context, then that context will resolve signal references this way.

STIL Conventions

Care must be exercised when defining multiple named SignalGroup blocks that may be used in different combinations with each other to resolve signal references in different contexts. If a signal group definition changes between two different contexts, then that group will have different definitions in at least two of these named SignalGroup blocks, and those two SignalGroups blocks cannot be used with each other.

4.11 WGL Timing Constructs

WGL timing information is defined very similarly to STIL; the corresponding WGL block to STIL Timing is called *timesets*, and the construct corresponding to the WaveformTable is the *timeplate*.

The biggest difference between timeplates and WaveformTables is the amount of information they each hold per Signal. Timeplate waveforms per Signal are limited to 2 drive values (essentially a '1' and a '0') and 2 compare values (also '1' and '0'), and a behavior for a Z operation, and an X behavior, while the STIL environment can support a much larger set of waveforms, using the array of WFC values available per WaveformTable. If a Signal uses multiple sets of event timings (multiple waveforms), WGL requires multiple timeplates to support this information.

When multiple Signals use multiple sets of events, the combination of different waveforms across all the Signals may require many timeplates to contain all the relationships. This can cause a combinatorial explosion of timeplate definitions in WGL. This same combinatorial explosion will also occur in STIL if the number of waveforms supported per Signal is constrained to a limited number as in WGL.

Another key differentiation between WGL waveforms and STIL waveforms is the explicit requirement in WGL to express an event behavior at time-zero in each waveform. Often this event is represented by the 'P' state in WGL, which has a similar definition to the STIL 'P' state. In fact, the STIL 'P' state was defined to facilitate interchange between WGL and STIL formats.

CHAPTER 5

Patterns

Standard
Test Interface
Language

The Patterns section defines the operation of the test, by repeated application of Waveforms on each Signal.

The application of waveforms across all signals for a unit of time is called a Vector in STIL terminology. A sequence of Vectors defines a STIL Pattern.

Pattern contents are extremely order sensitive; the Vectors contained in a Pattern will accomplish a test only when executed in the specified sequential order they are specified in.

5.1 Fundamental Pattern Constructs

The container for STIL Vectors is the Pattern block. A Pattern block may contain the following types of statements:

```
(5.1) Pattern pat_name {
(5.2)     W | WaveformTable wft_name ;
(5.3)     V | Vector { vector_statements }
(5.4)     C | Condition { vector_statements }
(5.5)     Call proc_name { vector_statements }
```

```
(5.6)    Macro macro_name { vector_statements }
(5.7)    Loop integer_count { pattern_statements }
(5.8)    MatchLoop integer_count { pattern_statements }
(5.9)    Goto label ;
(5.10)   BreakPoint { pattern_statements }
(5.11)   IDDQTestPoint ;
(5.12)   Stop ;
(5.13)   ScanChain chain_name ;
(5.14)}          // end Pattern
```

The **Call** and **Macro** statements shown above are thoroughly discussed in Chapter 11, Procedures and Macros. Likewise, the ScanChain statement is discussed in Chapter 12, STIL Scan.

5.2 Statement Order Dependency

Unlike the other STIL sections presented so far, the statements in the Pattern block are very order dependent. A test sequence relies on the proper sequence of data application in order to have the expected data response, and the test is organized by the sequence of Vector statements. In most cases two Vectors cannot be swapped and get the same device response. In addition, certain characteristics of the Vector statements also cause sensitivities with ordering; unexpected behavior may result due to the constructs present in Vector statements themselves. The most crucial of these behaviors is presented in Section 5.7.

Few STIL blocks have this level of internal statement order dependency. The Macros and Procedures block, which also contain Vector statements, have this same ordering dependency. Only these blocks have such a complete dependency on statement ordering.

This issue of statement ordering can be used to benefit. Many applications only need to process one Vector or a limited range of Vectors at a time. Each Vector represents all the activity for a segment of time, and processing of that time unit can be accomplished as a unit and the next unit (Vector) subsequently processed. It is important to remember why this capability is important: Pattern blocks may be very large, potentially containing millions of Vectors. It is not necessary to maintain all of these Vectors resident in memory for processing all at once, by taking advantage of the sequential behavior of this data and processing each Vector incrementally.

5.3 WaveformTable Reference Statement

A WaveformTable reference statement must be present before any Vector statements in the sequence of statements contained in a Pattern block, to identify the effective waveforms associated with the WFCs assigned to the Signals in the first Vector.

The WaveformTable statement is used subsequently whenever a different WaveformTable (containing, presumably, different timing and possibly different WFCs) needs to be applied. The WaveformTable statement identifies the WaveformTable in effect for the next set of Pattern statements until the next WaveformTable statement.

The current WaveformTable defines a single Period value. This value defines the amount of time taken by the current Vector, before the next Vector starts to execute. All waveforms across all Signals operate to this single Period. An individual waveform may have events that continue into the next Vector (STIL also allows a waveform to contain negative events that occur before this Vector starts — but care should be exercised with this behavior), but the next Vector will start when the current Period has elapsed.

STIL Conventions

It is possible to define multiple WaveformTable references between two Vectors. The last WaveformTable statement seen in the sequence before the next Vector will be the one that is applied for that Vector (and all subsequent Vectors until the WaveformTable reference is changed again). WaveformTable reference statements are not cumulative and each successive statement replaces the previous one. If multiple WaveformTable references are present it is recommended that the unused ones be eliminated when possible.

5.4 Signal Assignment

The primary behavior in a test program is the assignment of waveforms to signals in a test program. Waveforms provide both the stimulus and response information that define the test operation.

As presented earlier, the number of events that an ATE can support or define in a single unit of operation is limited, and therefore a test program consists of a sequence of these events, each defining one unit of the test. Most commonly these individual units are not divisible at test and other test activities must occur around these units.

Waveforms are collected under WaveformTables, and WaveformTables define a single period of time for this unit of operation to apply. The current WaveformTable reference specified by the WaveformTable reference statement in the previous section defines the active WaveformTable.

Waveforms are referenced by the WaveformCharacter that is associated with each waveform definition. The basic STIL statement to associate a waveform to a signal takes the form of an assignment statement:

```
(5.15)Signal = WaveformCharacter ;
```

The WaveformCharacter reference is evaluated from the WaveformTable that is currently applied. Remember that Waveform definitions are organized per-Signal (or group), therefore the same WaveformCharacter on different Signals does not necessarily define the same waveform behavior for those Signals. It is important to remember to de-reference the WaveformCharacter/Waveform/Signal relationship when interpreting a STIL test, rather than assuming a waveform behavior because of the WFC.

Signal WaveformCharacter assignment has a variety of representational options. For instance, in addition to referencing Signals in the statement above, SignalGroups that are aliases or renames of Signals can also be used. This is where SignalGroups come into use. Additional constructs are presented in the next sections.

5.4.1 Multiple Signal Assignment

The same basic assignment statement can be used to apply WFCs to a group of Signals as well:

```
(5.16)SignalGroup = WaveformCharacterList ;
```

When a group of Signals is assigned in STIL, each Signal must have its own WFC value or values. Therefore, the WaveformCharacter value in the first assignment becomes a WaveformCharacterList when assigned to a group of Signals. For example:

```
(5.17)Signals { "in1" In; "in2" In; }
(5.18)SignalGroups { "all_" = '"in1"+"in2"'; }
(5.19)...
(5.20)Patterns "first" {
(5.21)    W ...;
(5.22)    V { "all_" = AB; }
(5.23)}
```

WaveformCharacterLists are a special construct in STIL. They consist of strings of WaveformCharacters, which can be arbitrarily long. However, they must still obey the STIL token length limitation of 1024 characters, and therefore they may contain arbitrary whitespace, including carriage returns and tabs for readability purposes, but do not require the STIL concatenation operator when whitespace is present. WaveformCharacterLists can only appear on the right side of a signal-assignment construct and therefore parsing these strings is contained to this context.

WaveformCharacterLists also support a variety of modifiers unique to these strings, as presented in Section 5.4.5.

STIL Conventions

Do **not** use the STIL concatenation operator ('.') between whitespace-separated segments of WaveformCharacterLists. The concatenation operator is not required here because the STIL language defines whitespace to be part of the construct of a WaveformCharacterList, and in fact expects whitespace to separate long sequences of WaveformCharacters. The concatenation operator provides the converse function because it explicitly concatenates strings. Do not use it here.

Because WaveformCharacterLists consist of an arbitrary sequence of WaveformCharacters and are separated by arbitrary whitespace, these sequences may be constructed to appear identical to reserved STIL keywords. STIL keyword checks cannot be performed when processing WaveformCharacterLists because, in this context, these tokens are not keywords but are arbitrary strings of WaveformCharacters.

5.4.2 Multiple Based Signal Assignment

WaveformCharacterLists are arbitrarily long. But the contents of these strings are commonly limited to only a handful of different characters, often due to ATE environment constraints on the number of waveforms. A mechanism to compact this repeating data can drastically affect the overall size of a test program if long WaveformCharacterLists are common — and benefit transportability of the test by supporting smaller files.

STIL supports a mechanism to compact these character strings to a hexadecimal or decimal value, and identify the relationship of hex (or decimal) values to specific WaveformCharacters. There are several ways that this compaction is supported. First we will present the definition from the SignalGroups construct.

One of the SignalGroup attributes identified in Table 3.3 is Base. The Base attribute is used to establish a default interpretation of WaveformCharacters from a "Based" number (the term "Base" is derived from the context that the numbers being used to interpret WaveformCharacter values are either base-10 (decimal) or base-16 (hexadecimal) numbers). The syntax of this attribute is:

```
(5.24)Base (Hex | Dec) WaveformCharacterList ;
```

Here's how it works. Let us consider hexadecimal mapping first. The Base statement defines a mapping of WaveformCharacters, specified in the WaveformCharacterList, to discrete hex values. This mapping is dependent on the order of WFCs in the WaveformCharacterList; the first WFC is assigned the hex value 0, the second WFC is associated with the hex value 1, all the way up through all WFCs defined in this list.

To interpret a hex value assigned to this group, two distinct operations are performed. First, the hex value is partitioned by the number of bits necessary to support identifying each WFC in this mapping. This occurs implicitly in the mapping operation defined above — the number of bits is the smallest integral power-of-2 that contains the number of WaveformCharacters in the list. The second operation is that each set of bits is then equated to a hex value that identifies the index into this list and defines the WFC associated with this value.

Maybe an example will help clarify.

```
SignalGroups {
    "mapped" = '"in1" + "in2" + "in3" + "in4"' {
        Base Hex AB;
    }
}
...
Patterns "first_" {
    W ...
    V { "mapped" = 1; }
}
```

First, establish the mapping. The Base statement above specifies two WaveformCharacters in the list — A, and B. A is the first WFC, so it is assigned the value 0. B is the second WFC in the list, and is assigned the value 1.

Now, establish the number of bits necessary to support this definition. There are two WaveformCharacters in the list, and the smallest power-of-2 that supports the value ‘2’, is 1. Therefore only one bit is needed per Signal to map the hex values back to WFCs.

Now we are ready to decode the assignment statement. When the SignalGroup "mapped" is used in the Pattern block, the Base attribute defined on this group establishes how to interpret the assigned value. It is no longer a simple WaveformCharacterList but is now identified to be a hex value. A single hex value contains 4 bits, the value '1' above is equivalent to the binary representation 0001. Following the order of Signals in the group, "in1" is assigned WFC A, "in2" is assigned WFC A, "in3" is assigned WFC A, and "in4" gets the '1' value and is assigned WFC B.

What would happen if the group did not contain an integral factor of 4 Signals, for example, 3, or 5 Signals? In this situation we rely on an additional attribute to identify how to interpret the binary representation of the number. This is the Alignment attribute:

```
(5.35)Alignment < MSB | LSB > ;
```

The alignment attributes specifies whether the MSB of the binary representation, or the LSB of the binary representation, needs to be maintained while interpreting this assignment. This attribute **does not** change the order of the interpretation; neither the SignalGroup order nor the order of bits assigned to this group is modified by this attribute. Only the *starting index* in the binary representation of this number is affected by this attribute. When the LSB is specified, the lowest bits in the number will be maintained and the high order bits (if any) will be ignored. When the MSB is specified, the highest significant bit of the representation will be maintained, which for hex mapping means starting from the high order 4th bit of the first hex digit.

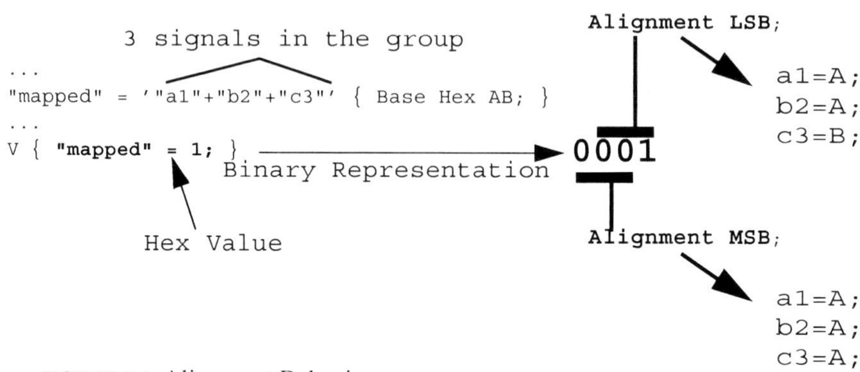

FIGURE 5.1 Alignment Behavior

5.4.3 Decimal Mapping

Decimal mapping is performed the same way as hex mapping, through the binary representation of the decimal number. The major difference with decimal mapping is that Alignment is not meaningful because decimal places do not correlate to binary indexes; decimal numbers are always processed from the LSB and zero extended as necessary on the MSB side of the value (Signal WFC assignment is always from positive based numbers). Also, decimal values cannot contain whitespace, unlike WaveformCharacters and hexadecimal values.

5.4.4 Default Attributes

All groups are defined with a default behavior that can be equated to a default attribute value. This is the value (or behavior) of this group when an attribute is not specified on the group. In particular, when the Base attribute is not specified, a SignalGroup receives a WaveformCharacterList that is interpreted to be a sequence of WaveformCharacters. When the Alignment attribute is not specified, the default interpretation of a number is MSB — to apply the most-significant bits and disregard any bits on the low side of the number that cannot be applied given the number of Signals in the reference. MSB was chosen as a default to support representation of scan data, which is presented later.

Signalgroups-on-the-fly only have default attribute values applied to them, no matter what attributes may have been defined on the members of the group. As presented earlier, attributes do not propagate up from members into a new group, and only the test attributes propagate down (from groups to the Signals).

5.4.5 WFC Expressions

We have just seen that WaveformCharacterLists are not just strings of WaveformCharacters, but may also be hex or decimal values. In addition, WaveformCharacterLists support several distinct operators specific for WaveformCharacterList manipulation. These operators are:

\r*integer_value*: specifies a repeat value for the next whitespace-separated word to be repeated. This is a textual repeat operation; no interpretation of the data being repeated is made until after the repeat has expanded the string. Repeats are convenient when the same WFC is being applied across a collection of Signals, or even when there is a pattern to the data being applied. The repeat specifies an integer number of repeats, followed by a whitespace-separated segment of characters to repeat. The integer value is specified without any separation from the **\r**. For example (assuming the 2 SignalGroups below do not define a Base attribute):

```
(5.36)"16bit_outbus"= \r16 X;// defines XXXXXXXXXXXXXXXX
(5.37)"16bit_inbus"=\r4 0110;// defines 0110011001100110
```

\w: the next set of characters in the WaveformCharacterList are interpreted directly as WaveformCharacters. This option will override any Base attribute specified for a SignalGroup and can be used to suppress a Base option. It can also be used in conjunction with the next several options, to change interpretation of a single WaveformCharacterList while the list is being read.

\h: the next set of characters in the WaveformCharacterList are interpreted as hexadecimal values. A Base attribute must have been present on the SignalGroup declaration, and the Base mapping defined in that attribute will be applied for the next set of characters. When this option is used, whitespace is required to separate this option from the set of hexadecimal values.

\h*WFClist*: the next set of characters in the WaveformCharacterList are interpreted as hexadecimal values, but the mapping values are specified here. A Base attribute is not necessary when this option is used (or is ignored if the attribute is present). This option can be used to compress WaveformCharacter data based on the specific WaveformCharacters used in the data. When a WaveformCharacterList is specified with a **\h** option, there cannot be whitespace between the **\h** and the WaveformCharacterList, otherwise the list would be interpreted as hexadecimal values.

\d, **\d***WFClist*: the next set of characters in the WaveformCharacterList are interpreted as decimal values, following the definitions above but using decimal values instead of hexadecimal values.

`\l`*integer_value*: defines a length to the data. This may be used to specify the number of WaveformCharacters necessary to be applied to SignalGroup, or indicate the need to truncate excess data. The length specification is useful in some data-dependent contexts such as scan, where there is no explicit definition of the data length. If present, this modifier can appear only once and at the beginning of the WaveformCharacterList.

STIL Conventions

Avoid using the `\l` modifier except in well-defined contexts. The STIL standard indicates that this modifier has two contexts of application, but does not define them clearly. Specifying the length of data is not necessary generally as the Alignment attribute specifies how to apply the data. Unless your application requires identification of the length of data externally, as in some scan contexts, this modifier is unnecessary.

5.4.6 Mixing Modifiers

STIL allows the WaveformCharacterList modifiers to be mixed in a single WaveformCharacterExpression. The fundamental rules for evaluating modifiers are:

- The repeat operator is evaluated at an innermost level. It operates over the next whitespace-separated set of characters and is expanded before interpretation of those characters as a Based value.
- After repeat expansion, the string is interpreted for Base representation in sequential fashion, starting with the default interpretation from the Base specified for this SignalGroup or Signal reference (and using WaveformCharacters if no Base is specified), and then changing to the specified Base as each `\h`, `\d`, and `\w` modifier is seen in sequence.
- After WaveformCharacter expansion, the `\l` modifier is applied.

5.5 Vectors

A Vector statement defines a set of waveforms to be applied against the Signals pertinent to the current test operation. Waveforms provide both the stimulus and response data. The waveforms are assigned to the Signals using the WaveformCharacter assignment construct in the previous section. An example of the start of a test program is shown below:

```
(5.38)Pattern "first_" {
(5.39)   W "default";// required initial WaveformTable
(5.40)   V { "allsigs_"=0101000XXXXX0000; }
(5.41)   ...
(5.42)}
```

The first Vector statement in a Pattern block needs to have a WaveformCharacter assigned for *every* Signal used in that Pattern. There is no default state or waveform assigned to Signals that are not assigned a WFC in the first Vector, although unused (*completely* unused) Signals may have a **DefaultState** value as defined in "Application of the Test Attributes" on page 53, Section 3.7.1.

After the first Vector, subsequent Vectors only need to contain signal assignments that change to a different waveform in that Vector. The incremental representation is discussed in Section 5.7.

Be careful when evaluating WaveformCharacter assignments during STIL processing. The current value of a WaveformCharacter is only valid at the time that the WaveformCharacter is actually executed on a Signal, in a Vector statement. Between Vectors, many things can change, including the WaveformTable in effect, and in some situations, transient WaveformCharacter assignments from the statement presented in the next section. Only the value present at the Vector statement is pertinent; transient values are ignored and STIL processors should not generate messages from these transient states.

5.6 Conditions

In addition to the Vector statement, STIL supports a Condition statement. This statement does not generate actual test events, nor does it contribute any test execution time. It is used only to establish new waveform assignments to Signals that may be applied at the next Vector statement.

Condition statements can appear any number of times before a Vector statement. Each Condition is processed separately, in linear sequence, and the resulting values of WFCs across all active Signals are applied to the next Condition or Vector statement.

A WFC can be assigned to a signal only once in a Vector or Condition statement (it may actually be assigned multiple times as long as all assignments are to the same WFC), but multiple Condition statements before a Vector may each assign different WFCs to a Signal — the last one before the Vector is the one that is carried into the Vector itself.

The following example demonstrates the use of the Condition statement to define the state of all Signals before the first Vector in a Pattern block:

```
(5.43)Pattern first {
(5.44)    W "default";// required initial WaveformTable
(5.45)    C { "allsigs_"=0101000XXXXX0000; }
(5.46)    V {}  // first Vector; values defined in the C.
(5.47)    ...
(5.48)}
```

Observe in the example above, that the first Vector statement is empty. This situation is allowed in STIL because all the Signal values were specified from the Condition statement that preceded the Vector statement. The Vector is responsible for asserting these values and the cycle/period time of the active WaveformTable. The previously defined WFC values for each Signal, specified from the Condition, are carried into this "empty" Vector and are applied at this point. This is an example of STIL incremental Vector specifications.

5.7 Incremental Behavior

After WFCs have been specified for all signals *by* the first Vector (*by* because the WFCs may have been specified from a Condition statement before the Vector as we have just seen), subsequent Vectors only need to present *changed* WFCs on Signals. The last specified WFC assignment to a Signal persists on that Signal until it is changed by a subsequent Condition or Vector statement. Consider the following example, which contains a single Signal assigned to a P WFC, and assume the P WFC represents a pulsed waveform:

```
(5.49) Pattern second {
(5.50)    W "default";
(5.51)    V { "onesig_"=P; }

(5.52)    V {}

(5.53)    V {}
(5.54) }
```

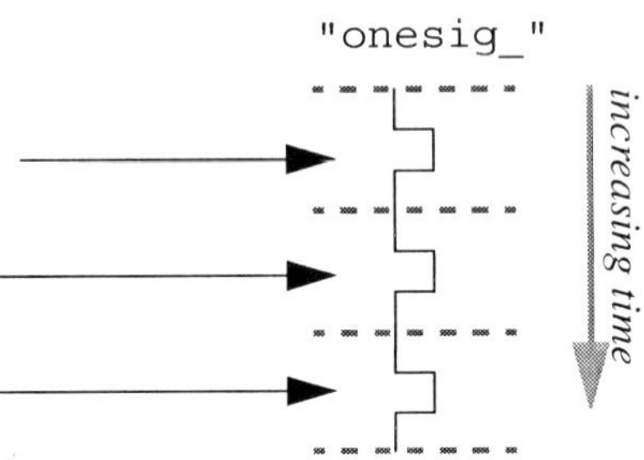

Note how the first Vector specifies the WFC P (and the pulsed waveform) for signal "onesig_", and that the next two Vectors do not change this assignment. The P remains assigned on this Signal, which causes the waveform associated with the P WFC to be re-asserted in these subsequent Vectors even though the P is not explicitly stated. This can be considered a “sticky state” behavior; the last assigned WFC to a Signal is “sticky” until changed to another value. The waveform diagram next to these statements represents the generated pulse from these statements, running in a vertical fashion with the Vector statements.

It is a common mistake to assume that the incremental behavior of STIL causes the last event *state* (i.e., DriveUp, etc.) specified in the last active waveform to remain on the Signal until changed. This is not the case. It is the last WFC assignment that remains on the Signal, and therefore the entire waveform behavior associated with that Signal is re-asserted on the Signal until the WFC value is changed.

There is another aspect of the incremental behavior that needs to be highlighted. Changing the current WaveformTable in effect can affect the current waveform on a Signal even with no other explicit changes to that Signal. To demonstrate this let us continue with the example we started above, but define two WaveformTable this time and a change to the WaveformTable during the Pattern:

```
Timing {
   WaveformTable "A" {
      Period '90ns';
      Waveforms {
         "onesig_" {
            P { '0ns' D; '30ns' U; '30ns' D; }}}}
   WaveformTable "B" {
      Period '90ns';
      Waveforms {
         "onesig_" {
            P { '0ns' D; }}}}}

Pattern second {
   W "A";
   V { "onesig_"=P; }

   V {}
   W "B"
   V {}
}
```

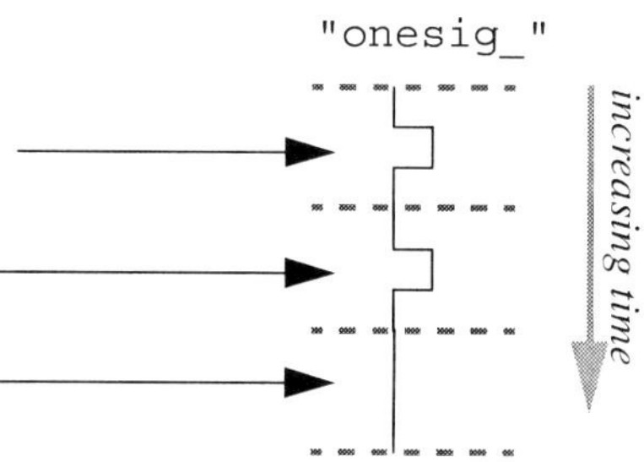

See how the Signal changed from the pulsed behavior defined in WaveformTable "A", to a low-value waveform without a pulse defined in WaveformTable "B", just by changing the active WaveformTable. This change occurred with no explicit changes on the Signal itself, but through the change in waveforms bound to the WFC P definition in each WaveformTable.

5.8 Loop Constructs

Test information is often full of repeating data. The fundamental cyclization process itself contributes to a basic level of repeatability in order to put the test into an ATE environment. STIL supports several constructs that allow sections of Pattern blocks to be identified as repeating. Being able to loop over these repeated test segments can be beneficial for several aspects of test processing. Using loops reduces the overall physical size of the test data (e.g., test file size), contributing directly to transportability (we can discuss separately the concern that using loops may affect the ability of the ATE to support the test data). And for ATE environments that support loop behaviors, the test program takes up less ATE memory allowing more test data to be defined.

5.8.1 Loop Statement

The simplest repeating construct in STIL is the Loop. A Loop is a block statement that contains additional statements found in the Pattern block, and has a defined single integer number of times specified to repeat these statements.

Many ATE environments differentiate between a "repeat" of a single Vector statement, and a "loop" around several statements. STIL does not differentiate these behaviors with separate statements. The same statement is used for both contexts. The ATE environment is responsible for identifying how (or whether) the STIL Loop can be supported.

A STIL Loop has a fixed number of iterations, hard-coded in the integer value associated with the Loop.

Loops may contain any other STIL Pattern statement, including most notably, other Loops. Again, STIL defines no restrictions on Loop count (except that it be a positive integer) nor on levels of nesting of Loops. The ATE environment often has restrictions on loop counts, and nesting. But satisfying the ATE restrictions can often be accomplished by a very straightforward transformation of the STIL data, as is discussed later.

A fragment of a STIL test with some representative Loops is presented next:

```
W "default";
V { CLK=P; outs=XXXX; }
Loop 10 { V {} }// A simple Repeat 10 times
                // of the previous Vector

Loop 21 {// A more complex Loop with multiple Vs
    W "new";
    V { CLK=P; }
    V { CLK=0; }
} // end Loop
```

5.8.2 MatchLoop

The MatchLoop construct is used in ATE environments to synchronize uninitializable elements in a design. It does this by looping over a sequence of Vectors that will eventually result in a deterministic state of the design but with an undefined number of loops to arrive at that state. Most ATE environments have special features to support this activity.

The MatchLoop provides this function by monitoring the state of specified outputs as part of the Vector information, defined by the Compare events on outputs. As long as the current response state does **not** match the specified state of all the Vectors in the MatchLoop, the loop will continue iterating. Implicitly, this operation is performing a "loop-until-pass" operation because the outputs have not yet matched and are therefore are failing this test from a normal test operation perspective. When the specified states on all Vectors in the MatchLoop match, only then will the MatchLoop exit at the end of the loop and continue the test execution from this deterministic state.

When a MatchLoop has matched, exiting the MatchLoop and resuming the main pattern sequence may take some time, depending on the ATE environment. To support re-synchronization with the flow of the main test program, MatchLoops require a BreakPoint construct to be present. BreakPoints support device operation to address issues like keep-alive clocks during the re-synchronization process. BreakPoints are discussed in Section 5.11.

MatchLoops should be used carefully. A failing device (or a poor test definition) may never satisfy the MatchLoop exit conditions, creating an infinite loop during test operation. Generally a time-out condition should be present, and it is also good test practice to validate the exit condition after the MatchLoop completes as well to make sure that the device is in the expected state. The MatchLoop supports one style of time-out designation, which is an integer loop count to identify the maximum number of loops that are required to synchronize the device. If this loop count is exceeded the test program will terminate. STIL also supports defining this count with the identifier **`Infinite`**, which identifies that the loop is not terminated by a count — but hopefully by some time-out someplace else.

MatchLoop operation can be very sensitive to ATE environments. Many ATE environments support a MatchLoop behavior, but vary on the details of how the output-checks are done, and on how (sometimes when) the loop is exited, due to pipeline or sequential processing of data during the test execution.

5.8.3 Other Loops

Inside Procedures and Macro blocks, an additional type of loop statement is supported, called the **`Shift`** block. Shift blocks support data dependent looping. This is presented in Chapter 12, STIL Scan.

The STIL Design Extensions define an additional Loop construct called Loop Data.

5.9 Labels

All statements in the Pattern block can be preceded by an optional label, using the following syntax:

```
(5.84) label :
```

The label text follows standard STIL username restrictions. These labels provide a method to tag each statement in the Pattern block with a message meaningful to some external observer to the test. Generally, labels are placed only at particular points of the test, for example when a device enters a different mode of operation.

Labels also serve an additional purpose. They provide a mechanism to reference a location in a STIL test. This location may then be used to jump into and out of a sequence of tests. As part of the pattern statements, STIL also defines the statement:

```
(5.85)Goto label ;
```

This is one mechanism that can jump around a section of STIL data (beware of jumping backwards). The PatternBurst also contains Start and Stop statements that reference labels, described in Chapter 6, Tying It All Together.

Because labels are used to identify locations in a STIL test, labels need to be unique in a test program. Labels are scoped to Pattern blocks, and to individual Procedure bodies (even though multiple invocations of a procedure results in multiple occurrences of the same label), and Labels are required to be unique inside these contexts. Because Macros are integrated into a calling context, labels in Macro bodies are ignored during STIL processing; in particular they cannot be used to resolve Goto statements.

STIL limits the ability to jump to labels as well. In general, Gotos are supported across the statements contained in a Pattern block or within a Procedure body, but jumping into or out of a Procedure during test execution is not allowed. Also jumping into or out of any Loop operation is identified in the STIL specification as resulting in undefined behavior. Don't jump in or out of loops.

The STIL Design Extensions and Tester Constraints extensions define additional labeling strategies as well.

STIL Conventions

Use labels conservatively. Because labels provide support for goto operations, label locations may need to be preserved to support these operations, and preserving long label text strings or many distinct labels during STIL processing adds overhead to the processing operations.

Avoid the use of Goto statements. This is not a particular STIL issue, but good coding practice and I had to say it here. Gotos promote unstructured test flow and make it difficult to track test execution. Also, jumping backwards in a test sequence (intentionally or otherwise) may create an infinite test loop that may not be exited easily. The intent of the Goto statement as part of the standard, is to support test debug and test development operations, with the expectation that normal test execution would not need or contain Goto statements.

5.10 Stop and IDDQTestPoint

The Stop and IDDQTestPoint statements can be considered labels with special purposes. They identify locations in the test program where something happens. There are no statement-specific values associated with these statements, just the individual keyword:

```
(5.86)    IDDQTestPoint ;
(5.87)    Stop ;
```

The IDDQTestPoint statement identifies a location in the test program where an IDDQ current measurement test may be performed. The mechanism to perform this test is not identified, nor are the consequences of performing the test. This is simply identified as a potential location for performing this test.

The `Stop` statement terminates execution of the pattern at this point. The intent of the `Stop` statement is to provide a debugging hook during test development; a final test program is not expected to contain `Stop` statements.

5.11 BreakPoint

The BreakPoint statement is one of the few statements in the STIL specification that identifies both the single statement and the block statement form, due to the different semantics of the two forms of this statement.

```
(5.88)    BreakPoint { pattern_statements }
```

```
(5.89)    BreakPoint ;
```

The purpose of the BreakPoint statement is to identify a point in the pattern execution where the device is in a stable state, where the main pattern execution may be momentarily halted while ATE-specific operations may be performed.

The two forms of the BreakPoint statement support different device operations during this BreakPoint execution. The simple statement indicates that the device has no operational requirements during this time. Holding all Signals at the last defined level or event state is sufficient to be able to resume the test at some later time. The block statement contains additional Vectors that need to be applied to the device during the BreakPoint time, for instance keep-alive clocks or other dynamic circuitry that needs active refresh operations.

When the block form containing multiple statements is used, the contained statements may be executed zero or more times. However, once a pass through the block is started, all pattern statements in that block will be executed. This is similar to the Loop behavior as the block is only exited at the end of the statements.

BreakPoints are somewhat of a quandary. They identify a point of stability in the device operation, but they are only required if the ATE environment needs to perform some operation at (or near) that point. Identifying *when* a BreakPoint will be needed is a function of the ATE environment; identifying *where* the BreakPoint may be placed in a test program is a function of the device behavior. Please review your application context to determine if BreakPoints are necessary in your STIL context or not.

5.12 WGL Pattern Constructs

WGL places all vectors inside a *patterns* construct, very similar to the STIL Pattern block, but a WGL test may contain only one *patterns* construct, whereas the STIL file may contain a collection of (named) Pattern blocks.

This implies that the STIL environment supports constructs for organizing pattern data, not needed in WGL environments. This is covered in the next Chapter.

The WGL *vector* statement is very similar to the STIL statement, except that each WGL vector requires a timeplate reference that defines the timing and waveform information over all Signals. And *each* signal must be associated with a waveform in a WGL vector. There is no incremental specification option in WGL.

WGL files of similar format are generally larger than STIL files. The lack of an incremental Vector construct is generally the primary contributor to file-size differences between STIL and WGL, unless timeplate combinatorial explosions, identified in the previous Chapter, are also an issue for the test representation. Scan data representation is also a factor, but that will be discussed in detail in Chapter 12, STIL Scan.

WGL supports both a single vector *repeat* construct and a *loop* across multiple statements, very similar to the STIL Loop construct except for the explicit differentiation in WGL between a single vector Repeat and a multiple vector Loop. WGL does not define a MatchLoop construct.

WGL supports a *symbolic* block that may be used to represent sets of signal assignments (WaveformCharacters in STIL) as a mnemonic label, and then apply these labels in the vectors. STIL does not have an equivalent capability.

WGL does not have an equivalent construct for the STIL Condition. The Condition statement is primarily used to set up a set of Signals for application in the following STIL Vector. Since WGL is not incremental on the Signal information per vector this construct would not be of benefit.

WGL pattern flow is externally controlled, so WGL does not have internal control flow options in the pattern like the STIL BreakPoint, Goto, and Stop. WGL does not have the IddqTestPoint statement either.

The WGL *scan* statement fundamentally serves the same purpose as the STIL `Shift` statement, but the relationship ends there. Data processing between the two statements is very different; the WGL scan statement is dependent on several additional blocks of WGL information, including data relating to the scan structures to provide and interpret scan data. The STIL scan statement is processed with the data provided directly and requires no additional constructs. This is covered in Chapter 12, STIL Scan.

The contents of the vector constructs themselves, and the relationship of STIL WaveformCharacters to WGL vector data, is covered in Section 9.5.

CHAPTER 6 Tying It All Together

Standard Test Interface Language

STIL tests use several additional constructs to sequence Pattern executions. This Chapter introduces the last set of constructs necessary to define a complete minimal STIL test.

6.1 Basic STIL Information Flow

We have seen how Signals are assigned activity by association with Waveforms interacting on a Vector-by-Vector basis, via the WaveformCharacter references. This was all presented in the previous Chapters. Vectors are then organized sequentially in a Pattern block. There are two additional levels of STIL information to organize both Patterns and to associate the Timing information with these Patterns. These blocks are the PatternBurst and PatternExec.

Each PatternExec in a STIL test represents a complete executable unit of a STIL program. By referencing Timing and PatternBurst blocks, a complete sequence of test data is assembled as shown in Figure 6.1.

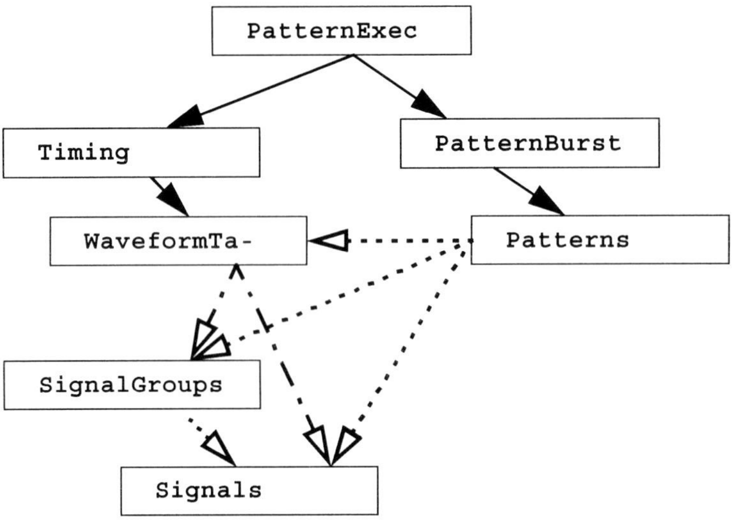

FIGURE 6.1 STIL Data Referencing

STIL data is resolved by combining both the Timing and the Patterns information, but as the dotted arrows in Figure 6.1, the actual Signal resolution between the Timing information and the Pattern information may be different. This is emphasized through the use of different dotted patterns for the two sets of references. Because the only reference to Timing information from the Pattern data is made by referencing named WaveformTables, by referencing different named Timing blocks (that contain the same-name WaveformTables), it is also possible to change the Timing on the same Pattern set. Both of these characteristics (independent Signal resolution between Timing and Pattern data, and independent Timing resolution) are important aspects of STIL tests and figure prominently in test reuse.

6.2 The PatternBurst

The PatternBurst organizes multiple Patterns and contexts of Pattern execution, to define a sequential execution that is expected to run in a single flow, or a burst. The syntax of the PatternBurst is:

```
(6.1) PatternBurst pat_burst_name {
(6.2)     (Environment/Context statements)
(6.3)     PatList {
(6.4)         pattern_or_burst_name {
(6.5)             (Environment/Context statements)
(6.6)         }  // end of Pat_name_or_Burst_name
(6.7)     }      // end of PatList
(6.8) }          // end of PatternBurst
```

The syntax above identifies the block form of the pattern_or_burst_name statements. This form of statement is only required when additional context statements are used. When just the pattern_or_burst_names are present, semicolon statements may be used for the statements in the PatList as well.

In its basic form, the PatternBurst contains a list of Patterns or other PatternBursts inside the PatList block. The Patterns are executed in the order they are presented in the PatList block. If the Patterns contain information that require named blocks to resolve, for instance a named SignalGroups block, then those blocks are referenced from the PatternBurst with the Environment/Context statements.

6.2.1 Basic PatternBurst

A STIL test program requires at least one PatternBurst to be complete. STIL Pattern blocks are organized only through the PatternBurst. The simplest PatternBurst contains a simple list of Pattern blocks:

```
PatternBurst "burst_one" {
    PatList {
        "pat_run_setup";
        "pat_functional";
    }            // end PatList
}
```

All PatternBursts must have a block name, and PatternBurst blocks must be referenced to be applied in a test. PatternBursts are referenced either from a PatternExec, or from other PatternBursts. When referenced from other PatternBursts, a hierarchical PatternBurst context is defined.

A name referenced from a PatList block is not intuitively a Pattern or PatternBurst name. However, as presented next, PatternBurst names must be defined before they are referenced from another PatternBurst. Therefore any name found in the PatList block, that was not previously defined as another PatternBurst, must be resolved as a Pattern reference. References to Patterns in the PatternBurst are expected to be forward-references; Pattern blocks are found at the end of a STIL test and are therefore defined after the references to them from the PatternBurst.

6.2.2 Hierarchical PatternBursts

STIL supports PatList blocks inside the PatternBurst referencing Patterns or other PatternBursts. This creates the opportunity for creating hierarchical constructs of PatternBursts referencing other PatternBursts. And this can support complex organizations of Pattern operation and reuse of Patterns with different environments.

When PatternBursts are referenced, the PatternBurst must be defined before it is referenced. This is required for two reasons: one, to eliminate the potential for circular references to PatternBursts, and two: to facilitate distinct identification of Pattern block names which must be forward references to the Pattern blocks found at the end of the STIL test.

Hierarchical PatternBursts support complex organizations of test data, and the reuse of Pattern blocks. Patterns may be reused in different contexts in the same test, through the use of different sets of other block referencing statements associated with the Pattern. These statements resolve all references in a particular Pattern by referencing different blocks the resolution for the same Pattern data may be different.

6.2.3 Environment/Context Statements in the PatternBurst

The Environment/Context statements referenced in Section 6.2 customize the execution environment of the PatternBurst, and include the following statements:

```
(6.9) SignalGroups signalgroups_block_name ;
(6.10)MacroDefs macrodefs_block_name ;
(6.11)Procedures procedures_block_name ;
(6.12)ScanStructures scanstructures_block_name ;
(6.13)Start label ;
(6.14)Stop label ;
(6.15)Termination { sigref_expr
(6.16)    ( TerminateHigh | TerminateLow |
(6.17)    TerminateOff | TerminateUnknown ) ; }
```

These context statements support the unique application of the same Pattern information, in the same STIL test, by resolving the information referenced inside the Pattern differently in the different contexts. Of course, if all the referencing statements are the same for two different PatternBurst references to a particular Pattern, then that Pattern is applied identically in both contexts.

The context statements may appear at two different locations inside the PatternBurst. They may be found outside the PatList block, at sort of a global level to the PatternBurst, or they may be found inside the block statement associated with each individual Pattern or other PatternBurst reference. When present outside the PatList block, each statement is applied to all the Patterns ultimately referenced by this PatternBurst. When defined inside a specific Pattern or PatternBurst block, these environment statements are associated with just that reference.

The primary statements that support mechanisms to resolve Pattern data differently in different contexts are the **SignalGroups**, **MacroDefs**, and **Procedures** reference statements. The **Termination** statements support a per-Pattern (or PatternBurst) specification of the Termination state as was presented in Section 3.6, and the restrictions identified on that attribute are applied in this context as well: when present, the Terminate statement can be specified only once on a Signal. The use of Termination as part of the Pattern-

Burst allows the Termination setting to be a dynamic characteristic of the test, changing (potentially) to different values during the test execution. **Start** and **Stop** statements control execution of a particular Pattern or sequence of Patterns, and the **ScanStructures** reference statement is covered later.

6.2.4 PatternBurst Organizations

Consider the next two sets of PatternBursts:

```
PatternBurst "pbA" {
    SignalGroups "first_group";
    MacroDefs "first_macros";
    Procedures "first_procs";
    PatList { "patA"; "patB"; "patC"; }
}
PatternBurst "pbB" {
    PatList { "pbA"; patD"; }
}
PatternBurst "pbC" {
    PatList { "pbB"; }
}
```

FIGURE 6.2 One Hierarchical Organization of PatternBursts

```
PatternBurst "pbA" {
    PatList { "patA"; "patB"; "patC"; }
}
PatternBurst "pbC" {
    PatList {
        "pbA" {
            SignalGroups "first_group";
            MacroDefs "first_macros";
            Procedures "first_procs"; }
        patD;
}
```

FIGURE 6.3 An Alternate Organization of PatternBursts

These two organizations provide exactly the same context of pattern sequence and pattern execution for the top level PatternBurst, which in these examples is named "pbC". Specifically both examples define the sequence of patterns and context of execution for each Pattern, from the definition of "pbC", as follows:

- Pattern "patA", with explicit SignalGroups "first_group", MacroDefs "first_macros", and Procedures "first_procs", and whatever implicit (unnamed) definitions might exist.
- Pattern "patB", with explicit SignalGroups "first_group", MacroDefs "first_macros", and Procedures "first_procs", and whatever implicit (unnamed) definitions might exist.
- Pattern "patC", with explicit SignalGroups "first_group", MacroDefs "first_macros", and Procedures "first_procs", and whatever implicit (unnamed) definitions might exist.
- Pattern "patD", with only implicit (unnamed) definitions that might exist.

Remember, you can always tell whether a PatternBurst references a Pattern or another PatternBurst inside the PatList statement, from the previous PatternBurst declarations since a PatternBurst must be declared before it can be referenced from another PatternBurst. This causes the highest level of a hierarchical set of PatternBurst definitions to be the last PatternBurst defined (or at least after any other PatternBurst that it references).

The bullets above emphasize the fundamental behavior of the context statements provided with each PatternBurst. Each context statement associated with a set of Patterns or PatternBursts becomes part of the execution environment associated with each individual Pattern that is ultimately run. Essentially the context statements accumulate for each Pattern that is referenced. It does not matter whether the context statements are associated with the Patterns from the top of the PatternBurst hierarchy, or from the bottom, or whether they are associated with all the PatList references (with the context statements outside the PatList), or whether they are associated with each individual reference using context statements inside the PatList, as long as they are associated with the set of Patterns you need.

So there are many options for expressing a sequence of Patterns in a hierarchical organization of PatternBursts.

These two examples help to identify a couple of recommended strategies for organizing Pattern references and their associated context information using

PatternBursts. As we have just seen, there are many options on where the context data may appear. Without an organizing strategy, it can be difficult to find where context information is located.

The primary strategy is to choose a location to place the environment statements, and be consistent once you have chosen a location. The two strategies that facilitate locating the environment statements, and reviewing the individual context of each Pattern, are to place the environment statements either at the lowest level of PatternBursts, or at the top level. Sometimes this is not effective; when a collection of Patterns are generated to operate as a single entity it may be more effective to have the environment, if it is shared by all the Patterns, be presented at an intermediate PatternBurst to maintain the association with all the common Patterns.

6.3 The PatternExec

The PatternExec is the head of a STIL test. A STIL file may contain multiple PatternExecs, following the standard STIL block naming conventions, and defining multiple STIL tests. Each PatternExec represents an independent test operation, at least at the level of the IEEE Std. 1450-1999 description.

The IEEE Std. 1450-1999 does not elaborate on how a PatternExec is run in a test environment, nor how multiple PatternExecs may be run. The actual execution of a STIL test is left to the test operation itself, and additional Work Group efforts.

The PatternExec performs one critical function for the STIL test definition. It binds the Timing information with PatternBursts. To perform this operation, both the Timing information referenced (if the Timing for this test is defined from named Timing blocks), and the PatternBurst referenced (remember all PatternBurst blocks are named) are referenced from the PatternExec block. In addition, the PatternExec may refine the time values through the use of some additional statements. The PatternBurst block contains the following STIL statements:

```
(6.18)PatternExec optional_patternexec_name {
(6.19)    Category category_name ;
(6.20)    Selector selector_name ;
(6.21)    Timing timing_name ;
(6.22)    PatternBurst patternburst_name ;
(6.23)}
```

The next section covers some additional STIL constructs that are not specific to the assembly of a STIL test, but are generic constructs and vital to the overall data that you may find in a STIL test.

6.4 Additional STIL Constructs

There are some additional STIL blocks required to define a complete, correct STIL test, and several that provide for additional description of the test information.

6.4.1 STIL block

As we said at the very start of Chapter 3, Signals, the very first statement in a STIL file is the STIL statement. The STIL statement may be either a single statement, referencing the primary standard, or it may be a block statement referencing the primary standard and additional extension standards. Both forms are shown next:

```
(6.24)STIL 1.0 ;   // Single Standard Reference
(6.25)STIL 1.0 {   // Multiple Standards
(6.26)    Design 2001 ;
(6.27)}
```

There are several other statement types, including comment statements, that may appear at any point in a STIL file, and arguably these statements may appear before a STIL statement. Hence the next convention.

STIL Conventions

It is strongly recommended that the STIL statement always appear as the very first statement in a STIL file. This allows certain types of file-checkers to validate STIL files as well as allows for quick visual inspection and identification of STIL files.

6.4.2 Header block

An optional block after the STIL statement is the Header block. Header blocks are allowed only after the STIL statement and before any other STIL block in that file; if present, the Header will appear at the top of the file after the STIL statement. Header blocks contain, fundamentally, file-oriented information although they may contain almost anything considered pertinent to the generator of this information. The defined statements of the Header block are:

```
(6.28)Header {
(6.29)   Title "title_string" ;
(6.30)   Date "date_string" ;
(6.31)   Source "source_string" ;
(6.32)   History { }
(6.33)}
```

STIL Conventions

Header blocks can appear at most once per file. There is no optional user name on this block since it can only appear once, although in a flow through multiple files (as described in the next Chapter) there may be multiple Header blocks seen. So remember that this block, even though unnamed, may be seen multiple times when reading one complete STIL test that is contained in several files.

All defined statements in the Header block do not identify the format of the information they may contain — *except for the* **Date** *statement*. The Date statement defines a very specific format for the information present, which in it's full form contains:

```
"Weekday Mon Day Hr:Min:Sec Timezone Year"
```

This must be enclosed in quotes to allow it to be a STIL string.

The History block of the Header defines no internal STIL statements. That means there is no standard mechanism for placing information in this block. While this would appear to be an oversight on the definition of the language, actually the Working Group consciously made this decision, using the capability of the Ann statement, defined next, to provide the basic mechanism for placing statements into this block, and attempting to maintain flexibility on the precise form of the data.

6.4.3 Annotations

STIL supports two forms of user-inserted comment information as part of the STIL constructs: Annotations, and Comments. While superficially providing similar functionality, the capabilities of each are different and there are distinct motivations for using one over the other. First we will present the Annotation.

Annotations are STIL statements. Actually, **Ann** is the STIL keyword, followed by special identifies to delimit the start and the end of a block of text that is the annotation body:

```
(6.34)Ann {* any text not containing *}
```

Ann statements are allowed to appear in STIL any place a STIL statement may appear. This allows the Ann statement to be associated with any part of STIL. The contents of an Ann statement are not defined and contain whatever information is considered necessary or useful. The only restriction on the contents of Ann statements is that they not contain a '*}', because that terminates the Ann body. The standard STIL token restrictions requiring whitespace also apply to the Ann body.

The freedom of location, and the flexibility of content, is a double-edged sword in the use of the Ann statement. Care should be exercised on the information maintained by STIL Ann statements. The principal application of the Ann statement is as a comment. Ann statements are not intended to maintain information that may affect test operation.

The Ann statement has one principal capability over the STIL comment construct, presented next. Ann statements are maintained through STIL processes, because they are part of the STIL information itself. The STIL standard defines that Ann statements, when present in a STIL test and read by a STIL tool, will be regenerated in any STIL output from that tool (for the section that contained the Ann statement). Of course it is dependent on the tool itself as to whether, or how, it may use the Ann statement itself.

Generally, to maintain association of an Ann statement with a part of STIL, the Ann statement should be placed inside the block construct of the STIL statement that the Ann is associated with. For example:

```
(6.35)STIL 1.0 {
(6.36)    DCLevels 2002;
(6.37)    Ann {*   only the DClevels construct
(6.38)             is used from the DCLevels spec *}
(6.39)}
```

The STIL Header History block, which contains no defined statements, often contains Ann statements as well:

```
(6.40)Header {
(6.41)    History {
(6.42)        Ann {* Netlist contained 3091 primitives,
(6.43)            1460 nets. *}
(6.44)        Ann {* Pattern block "A_1"
(6.45)            created Dec 2002 *}
(6.46)    }            // end History
(6.47)}                // end Header
```

STIL Conventions

Be careful with the contents, and the pertinence, of Ann statements. Ann statements are optimally used to hold comment data. But because of the flexibility of the placement of Ann statements, they may contain additional information as well, pertinent to the location of the statement. Rather than using Ann statements this way, consider at the options presented in Chapter 15, User Extensions.

The contents of an Ann statement (the Ann body) needs to be maintained intact in a STIL processing environment, to be able to regenerate the Ann statement in the meaningful context under which it originated. This implies most specifically that the original whitespace — not just the presence, but the actual characters, including carriage returns — needs to be maintained.

Ann statements are maintained with the STIL bock they are found inside, but processing operations (reading and rewriting files) may cause Ann statements to move within that block. The STIL standard does not require statement ordering to be maintained on most STIL blocks (except for certain blocks like the Pattern contents and the Pattern/PatternBurst references inside a PatList), and many STIL writers operate by placing statements in a specific order. When placing Ann statements into a STIL block consider whether the statements are still meaningful if they might be shifted about.

An alternate representation of one of the previous Ann examples is worth mentioning here. In the example at lines (6.35)-(6.39), an Ann was defined that pertained specifically to the DCLevels statement, which was presented as a single (semicolon terminated) STIL statement. An alternate representation that demonstrates the ability to support block statements, and the association of the Ann statement more closely to the construct it is referencing, is:

```
(6.48)STIL 1.0 {
(6.49)   DCLevels 2002 {
(6.50)      Ann {* only the DClevels construct is used *}
(6.51)   }      //end DCLevels
(6.52)}         //end STIL
```

6.4.4 Comments

STIL supports two types of syntax constructs for comments. These constructs are:

```
(6.53)// comment until the end of the line
(6.54)/* comment between these delimiters */
```

Comments are lexical constructs in STIL, which means they are part of the parsing operation before statements are recognized, and not a dedicated part of the STIL data flow. Comments are equivalent during STIL parsing to additional whitespace in a STIL file. They may appear any place that whitespace may appear in a STIL file, which means that comments can be placed in even more places in a STIL file than Ann statements. But comments are not maintained during STIL processing, and are typically discarded and replaced with whitespace if a STIL file is read and rewritten. Also, comments are not typically passed through applications and therefore may not be used by the application, either.

The two types of STIL comments serve different purposes. Line oriented comments are terminated with a carriage return, while the '`/*`' and '`*/`' is used either to create a comment block spanning several lines (with several carriage returns), or as a very small inter-line comment between other STIL statement tokens.

6.5 STIL Block Order

We have mentioned several times that STIL maintains a define-before-use mandate on most information. This places a general set of ordering requirements on the STIL blocks to satisfy these define-before-use restrictions. As we have presented the primary STIL blocks at this point, Figure 6.4 diagrams the primary expected order of STIL blocks in a complete STIL test.

Several blocks are not shown in this diagram. The STIL and Header blocks are not shown, because they appear as a function of the file structure of the STIL test. Some additional blocks that have not been presented yet are also not shown here,

There are several notable exceptions to the define-before-use rule. In particular, the Patterns blocks are found at the end of STIL test, and are definitely found after the PatternBursts. Remember the PatternBurst defines that any unresolved reference in the PatList is expected to be a forward-references to Patterns. Therefore any reference to a Patterns block in STIL, whether it is from a PatternBurst or some other construct, is most likely a forward-reference.

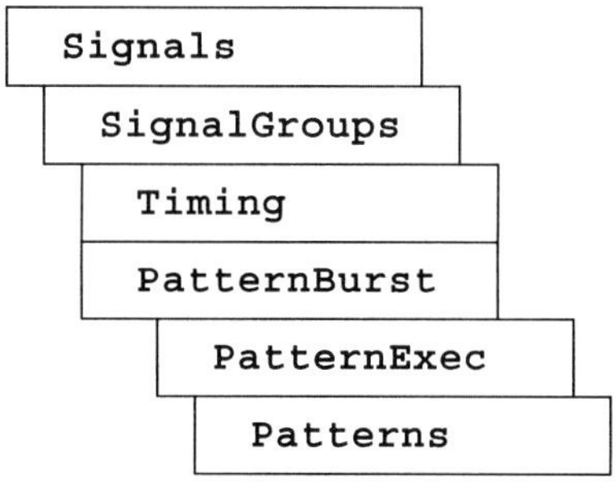

FIGURE 6.4 STIL Data Organization

6.6 WGL Constructs

The WGL environment does not support multiple patterns and does not define a resolution process dependent on referencing named sets of information. Each WGL test defines a single, absolute, complete entity. So the WGL environment does not have a corresponding construct to the STIL PatternBurst and PatternExec, and does not support a resolution process that may be dependent on other named blocks in the test.

CHAPTER 7

Files and the Include Statement

STIL Standard Test Interface Language

STIL supports a mechanism for partitioning STIL tests into multiple files. This construct, issues and considerations when applying this mechanism are presented here.

Whenever a test program is partitioned into multiple files, there is always a greater risk of loosing one file, and a corresponding higher maintenance cost for that test. However, if specific sections of information are generated from different sources, a flow that generates just those sections of information and then uses the Include to assemble all the information may result in an overall environment where the benefits outweigh the higher maintenance cost of multiple files.

7.1 Include Statement

The STIL Include statement is a language statement that may appear any place a STIL statement may be found. This is different than an include mechanism that is supported by a preprocessor environment (such as with the C language), where the include statement is handled as part of the lexical constructs of the language and may be placed at lexical, in addition to syntactic, boundaries.

But because the STIL Include statement may be found at any statement location there is still a lot of options about where it can be placed. And placing the Include on statement boundaries facilitates handling this statement as just another STIL construct, following the same STIL data model. And it avoids some of the ambiguity and complexity of preprocessed placement options.

The basic syntax of the Include statement is:

(7.1) **`Include`** *`filename`* `;`

There is an additional optional clause on this statement discussed in the next section.

The filename bears some consideration. The filename is treated as a STIL string or token (it is most likely enclosed in double quotes but need not be if it does not contain any special characters; that is rare since a filename almost always contains a period). This name is interpreted by the application, and resolving this name to a STIL file is a function of the application.

STIL Conventions

It is recommended to always quote filenames in the Include statement. Generally a filename will contain at least a period (.) character which has other semantic meanings in STIL, and often names may contain other characters as well only supported as quoted STIL strings. Uniformly quoting filenames reduces the potential for mistakes when creating STIL files.

It is recommended that top level STIL files are named with the suffix **`.stil`**. STIL files that are included may have a suffix appropriate to the contents of those files, for example .pat or .stil.pat. At the very least, a naming convention should be defined to facilitate identifying STIL files, and in particular, top level STIL files.

7.1.1 Relative File Path Naming

There are strategies to minimize the cost of maintaining multiple files, that impact the filenames that are present here. For example, a STIL test could consist of a directory of files; all files would be found below the top directory for this test. Transporting this set of files is straightforward because all files related to this test are found under this directory structure. All filenames present in Include statements would be best expressed as relative paths to each other, allowing relocation of the entire test set with no impact to the test itself. However, sharing information or maintaining common information between two sets of tests (e.g., of similar products) is not straightforward in this structure since each test would have it's own copy of information.

7.1.2 Absolute File Path Naming

Another common naming convention is the use of absolute file path names. The advantage of specifying a complete file path is that the location of the file is clearly represented. The disadvantage is a relatively higher "fragility" of the overall system; any changes to the environment are likely to force edits in the STIL files to maintain operation of this set of STIL files.

7.1.3 Using Logicals in the Path Name

Some of the fragility of a fully specified file path can be avoided by using system environment variables, also called logical variables, to represent parts of the file path name. Changing the location of the STIL files may cause a change to the environment variable definition but not necessitate individual file edits to change each Include reference. Environment variables are referenced as part of file names in system-dependent ways; one of the primary reasons the STIL standard does not define how to interpret file names is the handling of constructs like environment variables as part of the name.

The disadvantage with logicals is the need to maintain an external environment definition in order to read the STIL data. Over time these definitions have a tendency to be lost, and trying to resurrect a previous environment may not be easily accomplished.

7.2 The IfNeed Option

The Include statement has an optional clause to identify the *intended* contents of the included file. This clause is used by the application to determine if the information identified to be contained in this referenced file, is needed by the application or not. If the application does not use this information then it may chose to skip over this file completely.

The STIL Include statement with the IfNeed construct looks like:

(7.2) **Include** *filename* **IfNeed** *STILBlockName* ;

STIL Conventions

The IfNeed statement may be used by a STIL application to skip over a file that contains information, based on the name of the STIL block, that is not needed or used in that application. However, there are no requirements that the STIL file referenced by the Include actually contains this block of information. Also, the STIL file being Included (or skipped) may contain more than one type of STIL block. There should not be a problem with an application always processing the Included file, and it may be useful in the application to provide override options if the IfNeed is somehow incorrect for the purposes of that app

7.3 Additional File Constructs and Behaviors

Each STIL file is required to start with a **STIL** statement, and may be followed by a **Header** block. These two statements were presented in Section 6.4. It is important to remember that a STIL file, no matter what it contains, must have the **STIL** statement as the first statement in the file.

The Include statement is placed on STIL statement boundaries. This means that a STIL Include cannot start in the middle of a STIL statement, which prevents parsing a single STIL statement across two (or more) files. However, it is possible to split a STIL statement at the end of a file across two (or more) files, if one file just ends in the middle of a STIL statement and the file that

Included that file continues with the rest of that statement. But this is not recommended, if for no other reason than the maintenance issues caused by this situation. So for the next convention.

STIL Conventions

STIL files should contain complete STIL statements. The files may contain partial STIL blocks; a single STIL block may be split across multiple files, but individual STIL statements should not be split between multiple files. In particular, when a STIL file ends, it should end with a complete STIL file that ends with a '`;`' or '`}`'.

CHAPTER 8

Specifications

Standard Test Interface Language

Chapter 4, Timing, identified the fundamental constructs for timing representation in STIL tests. STIL supports several additional dimensions of flexibility in the definition of Timing constructs. One of these dimensions is the use of parameterized values for specifying the timing values. These constructs are presented here.

8.1 Device Operating Specifications

Devices are required to operate within specified limits. This is necessary to support integrating the device into a system — if the device operates outside of some expected range, then the device will not respond properly when it is plugged into the system.

These operational ranges become an additional aspect of the test program that needs to be verified. The device needs to be validated to be within these operational ranges in order to be qualified to use. These operational ranges cover the entire gamut of operating parameters: timing characteristics as well as other parametric behavior such as voltage levels and current drive capabilities. Not all of these parameters need to be directly tested on a per-unit basis.

Characterization information can be extracted from special die on a per-lot or per-wafer basis, and the information extracted from these die establish common characteristics across a common lot, a common wafer, or sometimes just that region of the wafer.

Because there is a specified range of acceptable behavior behind these parameters, STIL supports a mechanism to define expressions and constructs to represent these ranges.

Many device specifications are identified with three values: a minimum value, a maximum value, and a typical value. The minimum and maximum values identify the operating range over which this device is considered to be properly functional. The typical value represents the operating value that most devices are expected to be near. From a test perspective, the minimum and maximum values have the most interest as these are the limits that qualify valid device operation. The typical value is often used in design contexts, for example in simulation runs, to validate the device or system operation from a 'typical' perspective.

To maintain this context of device specification, STIL supports an expression for each of these values, and two selection constructs to identify which value to use during test operation. One selection mechanism identifies the "category" of the timing variables, the other selection mechanism identifies which of the 3 values to use for each Variable.

The STIL block **`Spec`** holds the specification expressions. Spec blocks are organized around Categories. Each **`Category`** block may define a minimum, typical, and maximum expression for each variable. The STIL block **`Selector`** identifies which of the minimum, typical, or maximum value to use for each variable.

8.2 STIL Extensions to the Spec Variables

The STIL specification (IEEE Std. 1450-1999) defines a range of unit types (Seconds, Volts, etc.). But the standard itself only supports one type of expression using time units, called a *time_expr* in the specification.

STIL extensions to support DC Levels (IEEE Std. 1450.2-2002) extends the application of these blocks to encompass DC parametric expression support as well. DC Levels are covered in Chapter 13, STIL Levels.

The presentation of Spec Variables in this Chapter is extended to include the additional expression types supported in DC Levels. This does not change the language constructs of this section but only allows the expression types to be more flexible in what they are able to represent. Again, for these spec variables definitions, the syntax constructs are the same, and all constructs used here are defined in the baseline STIL specification. The extension specifications simply provide extended utility to these definitions.

8.3 Spec and Category

The STIL `Spec` block is used to define spec variables and the expressions behind them. STIL `Spec` blocks, however, are really just containers for Category information. Spec blocks follow standard STIL naming conventions: one Spec block may be unnamed, all other blocks require a name that is different than any other Spec block name in the test. However, the similarity ends there.

All Spec blocks, whether named or unnamed, are treated as global definitions in STIL. Furthermore, there are no referencing mechanisms in STIL to the Spec block names themselves; Spec block names are never used in STIL. However, what is referenced are the Category names contained in the Spec blocks.

Spec blocks serve only as containers for Category blocks. Category blocks contain sets of Variables, and each Variable declaration has the option to define 3 expressions, associated with the minimum, maximum, or typical context of that expression.

Here is another place where these constructs differ from other STIL processing. For each Category name defined, all Variables of that Category are collected together in a single Category definition, no matter how the Category definitions were spread across Spec blocks. Variables must be unique in a single Category; a variable and expression-type (min, typ, max) can be defined only once in a Category. But it is expected, in fact required, that the same variables be defined in multiple Categories if these Categories are applied to the test independently. Each Variable declaration in a Category is independent of other Category declarations of that Variable — each Category/Variable/Type (min, typ, max) combination contains an independent and separate expression.

The Spec block syntax is also somewhat different than other STIL constructs. The syntax of the Spec block comes in two flavors: a Category-oriented flavor, where the Variables are defined inside the Category, and a Variable-oriented flavor, where the Categories are defined as part of the Variable. Typically, a STIL test would use only one or the other flavor of data representation. These flavors are purely representational options for the STIL data. From a data handling perspective, all Variables are always associated (and collected) under the defined Category.

Statements supported under the Spec block take the form:

```
(8.1) Spec optional_spec_block_name {
(8.2)    Category category_name {
(8.3)       spec_variable_name = based_expression ;
                  - or -
(8.4)       spec_variable_name {
(8.5)          Min based_expression;
(8.6)          Typ based_expression;
(8.7)          Max based_expression;
(8.8)       }
(8.9)    }        // end Category block

                  - or -
```

```
(8.10)    Variable spec_variable_name {
(8.11)        category_name = based_expression ;
                      — or —
(8.12)        category_name {
(8.13)            Min based_expression;
(8.14)            Typ based_expression;
(8.15)            Max based_expression;
(8.16)        }
(8.17)    }           // end Variable block
(8.18)}               // end Spec block
```

STIL Conventions

It is recommended to chose one form of the Spec statements, either the Category/Variable or the Variable/Category construct, and use it uniformly in your STIL test. While there is no restriction in STIL about mixing these two forms, human interpretation of the STIL test can be confused when both forms are present because of the similarity of the statements defined for both constructs.

There is no checking performed in STIL on the creation of Category names. Variables defined under the same Category will be merged into a single set of definitions when that Category is accessed. *Variables defined under a misspelled Category name may never be found.* If your STIL process complains about undefined Spec Variables, review the Category names present in the STIL test (particularly for the missing Spec Variables).

There is no unnamed Category block. Variables are only defined under named Categories. For that matter, there is no unnamed Variables, either.

For ease of maintenance, keep all types (min, typ, max) of a Variable collected under one location. Because STIL will merge Categories together, it is possible to distribute the multiple expressions for one Spec Variable in a variety of fashions, but it may not be easy to locate all the locations of one variable definition when this is done.

If a Spec Variable does not require multiple expressions for **min**, **typ**, or **max** description, consider using the assignment statement for this definition. The assignment statement will define the **typ** value of this variable only, and all applications of that variable will reference that single definition. If, however, multiple expressions are defined for this variable then it is recommended to use the block statement form the variable definition. As in the previous comment, there is no STIL restriction with defining the typ value from the assignment statement and then using the block construct to define the min or max, but readability and human interpretation are affected if these styles of variable declaration are mixed.

8.4 Spec Variable Resolution

Unlike SignalGroups (which can be redefined in different named blocks), Spec Variables can be defined only once for each Type and Category. Spec Variables cannot be redefined multiple times; there is no advantage to incremental definition and replacement of a Spec Variable like there is with SignalGroups. Because Spec Variables can be defined only once, the definition sequence of Spec Variables can be written in to define non-dependent Spec Variables first, and Spec Variables that depend on other Spec Variables later. Circular expression evaluation, where two (or more) Spec Variables definitions reference each other's Spec Variables, cannot be resolved and identify improper Spec Variable definitions.

Spec Variable expressions cannot be evaluated in general, in advance of expression selection. Each Spec Variable may define multiple types, and a single expression with multiple Spec Variables may have as many different values as combinations of variable types. Spec Variable selection, presented in Section 8.6, supports per-variable identification of the type; until each Spec Variable type to use is defined, a single expression value may not be identified.

8.5 Based Expressions

Spec Variables are assigned values from "based expressions". Based expressions are either simple, single constructs containing a value and a unit (with or without a scale factor), or are arithmetic expressions containing these quantities, references to other spec variables, and arithmetic operators.

We have already seen the simple expression construct defined for time applications, in Section 4.3. The basic time expression defined in that section is one example of the simple construct.

Arithmetic expressions support all the standard operators — plus ('+'), minus ('-'), division ('/') and multiplication ('*'), as well as parenthesis for grouping and comparison operators.

The following Spec example provides one application of Spec Variables:

```
Spec "1" {
    Category "full_voltage" {
        In_Z_time_ = '2ns';
        Out_tristate_delay_ = '10ns';
        Out_Z_time_ {
            Min 'In_Z_time_ + Out_tristate_delay_ - 2ns';
            Max 'In_Z_time_ + Out_tristate_delay_ + 3ns';
            Typ 'In_Z_time_ + Out_tristate_delay_';
        }
    }               // end Category "full_voltage"
    Category "threshold_voltage" {
        In_Z_time_ = '2ns';
        Out_tristate_delay_ = '15ns';
        Out_Z_time_ {
            Min 'In_Z_time_ + Out_tristate_delay_ - 4ns';
            Max 'In_Z_time_ + Out_tristate_delay_ + 7ns';
            Typ 'In_Z_time_ + Out_tristate_delay_';
        }
    }               // end Category "threshold_voltage"
}                   // end Spec "1"
```

8.6 Selector

There are two constructs to reference when a Spec Variable is used in a STIL test context: the Category (or Categories) to apply to this context, which identifies the set of Variables available, and the selected value (min, typ, max) of the Spec Variable to use.

The selection process requires a reference to the STIL **Selector** block. The **Selector** block has the following syntax:

```
(8.19)Selector selector_block_name {
(8.20)    spec_variable_name < Min | Typ | Max > ;
(8.21)}
```

The Selector construct is straightforward; it simply identifies for all variables that have multiple types, which type to use. To continue with the previous example and define a couple of Selectors for the one variable of the Spec block named "1" that needs selection:

```
Selector "min_times" { Out_Z_time_ Min; }
Selector "max_times" { Out_Z_time_ Max; }
```

STIL Conventions

All STIL Selector blocks have names. There is no unnamed (and globally accessible) Selector block. This was a conscious decision in the specification, and not just an attempt to complicate the language with yet another non-uniform behavior. The STIL development effort identified the requirement to explicitly resolve all multiple-valued variables, because it was anticipated that different Categories may not define (or need) all types of a specific Spec Variable for the context of that Category. Forcing definitions across all Categories, to support a global Selector that may never be used (or usable) in practice was seen to be inappropriate.

8.7 Applying Spec Variables

Lets finish that example we started in Section 8.5. There are two additional pieces of information needed to apply these Spec Variables in a STIL test.

The first additional behavior is the use of the Based Expressions defined in Section 8.5, as part of the Timing Expressions presented in Section 4.3. The time expression references in a Waveform support the Based Expression context, including full arithmetic expressions and referencing Spec Variables.

So here is an example of a Timing block using the Spec Variables we defined previously:

```
Timing { WaveformTable one {
    Period '100ns';
    Waveforms {
        input_Sig { A{ 'In_Z_time_' U; }}
        output_Sig { Z{ 'Out_Z_time_' T; }}
    }
}} // end Timing
```

The last construct is the reference to the Category and Selector. This is done from the PatternExec with the following reference statements:

(8.22) **Category** *category_block_name* ;
(8.23) **Selector** *selector_block_name* ;

So to complete the example:

```
PatternExec "A" {
    Category "full_voltage";
    Selector "min_times";
    PatternBurst "first";
}
PatternExec "B" {
    Category "threshold_voltage";
    Selector "max_times";
    PatternBurst "first";
}
```

Which defines two unique sets of timing, applied to the same Pattern set (in different bursts). PatternExec "A" contains the timing:

```
Timing { WaveformTable one {
    Period '100ns';
    Waveforms {
        input_Sig  { A { '2ns' U; }}
        output_Sig { Z { '10ns' T; }}
    }
}}      // end Timing
```

PatternExec "B" contains the timing:

```
Timing { WaveformTable one {
    Period '100ns';
    Waveforms {
        input_Sig  { A { '2ns' U; }}
        output_Sig { Z { '34ns' T; }}
    }
}}      // end Timing
```

8.8 More Spec Constructs

8.8.1 Explicit Spec Variable Type Referencing

STIL supports bypassing the Selector mechanism for Spec Variables with multiple types, by explicitly referencing the type using the construct:

(8.24) *spec_variable*.< **Min** | **Typ** | **Max** >

For example, **Out_Z_time_.Min**. This construct resolves the variable independently of any type values but still requires a Category to specify the value. This may be used to specify a different type in resolved variable context, or to prevent a variable value from changing between expressions.

8.8.2 The Spec Variable Meas Type

Spec Variables have a forth type besides **Min Typ** and **Max**. This is the **Meas** type, short for *measured.*

The **Meas** type is used to specify a value for a variable that is set from an external source. The STIL specification does not specify how a measured value is specified; it is expected to be set from some other test operation.

STIL Conventions

Meas values are not portable, by definition. The process to establish a measured value is not defined and must be resolved on a per-application basis.

8.9 Relative Waveform Timing

The WaveformTable block inside the Timing block supports additional constructs in the Waveform, to define event timing relative to other events in waveforms in that WaveformTable. These relative constructs are contained within the WaveformTable itself. The Spec and Selector mechanisms are not required to use relative timing, and in fact the Spec and Selector are excluded from leveraging these constructs (although these constructs in the WaveformTable can leverage spec variables from within the Waveform). These constructs are presented here because they are consistent with the notions of parameterized timing provided by Specs and Selectors.

Before we present these constructs, we need to augment the syntax definition of the Waveform event statement, introduced in statement (4.1) on page 60:

```
(8.25)event_label : 'time_expression' state_value(s);
```

The event statement in STIL supports an optional event_label, followed by a semicolon just like other STIL labels. This event label may serve to document the event statement in some way, or may be used directly as a spec variable in other time_expressions *within this WaveformTable*.

We also need to augment the definition of a time_expression. The time_expression was previously extended in Section 8.5 to support referencing spec variables in the expression. In addition to referencing spec variables, the time_expression statement *inside a WaveformTable* may reference event_labels and reference two new constructs to identify relative timing within the waveform itself - '@' and '@(integer)'.

The '@' construct by itself references the immediately-preceding event statement to the left of this statement. It operates like a specialized spec variable: the time value of the expression in the event statement to the left will be used here. The '@(integer)' construct references the (integer) event statement within this waveform, starting with the first event statement counted as '1'. So '@4' references the forth event statement in this waveform.

Figure 8.1 presents an example of using these relative time constructs. The first waveform defines an event_label. The second waveform references this event_label to define the timing of the first event, and then uses the '@' and '@1' constructs to reference this timing inside this waveform, for the next two event statements.

```
Timing { WaveformTable one {
    Period '100ns';
    Waveforms {
     inputs { 10 { start_one:'2ns' U/D; }}
     outputs{ HL { 'start_one + 10ns' Z; '@ + 80ns' h/l;
                          '@1 + 88ns' X; }}
    }
}}      // end Timing
```

FIGURE 8.1 Relative Timing Constructs

8.10 Styles of Timing Representation

Obviously from all the previous constructs, there can be several styles of timing representation in STIL. These styles can be summarized based on the constructs representing the time values in the Waveforms. These types of timing representation are categorized as follows:

- Fixed Timing, with single, constant time values for each event defined in the Waveforms. This is the most direct representation as presented in Chapter 4, Timing. This style is generated from most CAD environments that do not have detailed device timing information, for example Automatic Test Pattern Generation tools that rely on a timing-independent device model for pattern generation.

These tools depend primarily on an event order relationship but do not have specific timing information and rely on the timing to be externally provided (commonly as fixed values).

- Spec Variable-Dependent Timing. The waveform time events contain only single references to Spec Variables, and the resolution process for Spec Variables is used to arrive at the timing applied to the Waveforms. All expressions are defined in the Spec block only. This style of timing permits easier modification of time values across affected events in a WaveformTable, depending on the organization of the Spec Variables and the expressions used. At the same time, because the time expressions are not distinctly identified with the events they affect, it may be more difficult to identify the consequences of changing a single Spec Variable definition. Some amount of consideration for structuring the Spec Variable definitions is critical to support and maintain this type of timing.
- Expressions and Relative Timing defined in the Waveforms. This is an extension to Fixed Timing, where the waveforms specify time events that are expressions, possibly also using the '@'-timing options to specify relative events, but do not use Spec Variables. Again, depending on the use of the '@' operator, single changes to one timing value may affect multiple events — which may be the desired effect if the relationship of the events is proper as it reduces the number of edits to make a change, or it may result in unexpected additional effects if the relationships are not proper for that specific edit. Again, consideration on the proper structuring of the relationships is critical for optimal benefit.
- Relative Timing with Spec Variables. This style is the most complex, combining timing information from all locations. This combination relies on expressions present in both the Waveforms (to support '@'-relative events) and on Spec Variables, with expressions present in the Spec block as well.

The last style of timing definition is the most difficult to develop, but potentially has the most benefit from a maintenance or flexibility perspective if developed in a logical, structured manner. This is a subject of more detail than appropriate for this book. The point here is to be aware of these styles of timing, and aware of the implications of each.

8.11 WGL Specification Constructs

WGL supports a very flexible timing and levels expression environment through the *equationsheet* and *equationdefaults* constructs. These constructs support declaration of variables, definition of values through expressions, and multiple values (min/max) definitions. WGL also supports creation of multiple sets of values for variables with the *exprset* construct which is similar to the STIL Category. Mapping across these two environments can be a complex process.

CHAPTER 9

Partitioning Timing

STIL allows timing and corresponding pattern information to be organized in a continuum ranging between what we will call the All-Waveforms-in-One-Table organization, to the One-Waveform-Per-Table organization. There are benefits, and disadvantages, to each end of the continuum and to all points in between. This Chapter identifies the endpoints of the continuum, and the issues to consider in between.

How timing information is partitioned may be as much a personal preference as a mechanism for transmitting additional information about the way the timing is considered to be applied in the application context. There may be device-oriented reasons why the timing information is organized the way it is. But more often the organization is justified based on the test environment. Applying a set of organizational operatives on the timing information may lead to a direct method of identifying if a test program can operate on a particular ATE. We will identify some classes of these operatives here, and mechanisms of transforms to support these operatives.

We start with a presentation of the two extremes. Please remember: these next two examples present extreme behaviors of the language. *They are not intended to be practical.*

This chapter presents some issues with Timing in the STIL environment — how it can be described and how it ought to be described. No new language constructs are identified in this Chapter.

9.1 All-Waveforms-in-One-Table

As long as the number of Waveforms per Signal does not exceed the number of WaveformCharacters supported in STIL (which is 63 characters, upper- and lower-case alphabetic, numeric, and underscore), all timing on all Signals can be presented in a single WaveformTable. Additional WaveformTables are only necessary to support additional waveforms if they exceed the number supported by one WaveformTable. The example below presents this context. Note that this example has also grouped all signals together in a single All_Signals group reference:

```
Timing {
   WaveformTable wft_A {
      Waveforms {
         All_Signals {
            Aa { '0ns' U/D; }
            Bb { '10ns' U/D; }
            CcX { '0ns' Z; '50ns' H/L/X; }
         }}}}
```

Patterns that use this WaveformTable could look like:

```
Pattern "first" {
   W wft_A;
   V { All_Signals=AAAAAAAXXXX; }
   V { All_Signals=AAaAAABXXCX; }
   V { All_Signals=AaaaaABccCc; }
}
```

Because all Signals were collectively referenced in the group All_Signals, all Signals were allocated all WaveformCharacters. It is unlikely that all the Signals would use all WaveformCharacters in most situations; for example output signals would have no need of input-only waveforms. But perhaps more important is what input waveforms are used on what set or subset of inputs, and correspondingly, what output waveforms are used on what subset of outputs. Without reading through the Patterns there is no direct way to determine what Signals actually use what WaveformCharacters.

9.2 One-Waveform-Per-Table

The other end of the spectrum is to allow only one Waveform definition for each Signal per WaveformTable. Each time any Signal changes Waveforms causes a new WaveformTable to be defined. Because only one Waveform is defined for each Signal, the next example only uses one WaveformCharacter to represent that Waveform for each Signal — however remember that the waveform behind that WaveformCharacter may be different for each Signal:

```
Timing {
   WaveformTable wft_A {
      Waveforms {
         Some_Sigs {
            A { '0ns' U; }}
         Rest_of_Sigs {
            A { '0ns' Z; '50ns' X; }}
      }}
   WaveformTable wft_B {
      Waveforms {
         A_couple_Sigs {
            A { '0ns' U; }}
         A_couple_more_Sigs {
            A { '0ns' D; }}
         Other_Sigs {
            A { '0ns' Z; '50ns' X; }}
         More_other_Sigs {
            A { '10ns' U; }}
         Remaining_Sigs {
            A { '0ns' Z; '50ns' H; }}
      }}
   WaveformTable wft_C {
      Waveforms {
         A_couple_dif_Sigs {
            A { '0ns' U; }}
```

```
            A_couple_more_dif_Sigs {
               A { '0ns' D; }}
            Other_dif_Sigs {
               A { '0ns' Z; '50ns' L; }}
            More_other_dif_Sigs {
               A { '10ns' U; }}
            Remaining_dif_Sigs {
               A { '0ns' Z; '50ns' H; }}
         }}
}
```

Patterns that use this WaveformTable could look like:

```
Pattern "first" {
   W wft_A; V { All_Signals=AAAAAAAAAA; }
   W wft_B; V {}
   W wft_C; V {}
}
```

This example is incomplete without defining each of the SignalGroups referenced above, but the concept should be clear. Because only one WaveformCharacter is needed, the Pattern data becomes only a reference of the WaveformTable for each Vector. In this context the Pattern data is of minimal use. All of the information has been encoded into the WaveformTables applied at each Vector.

What is not clear (somewhat purposefully in this example) is what waveform is associated with each use of WaveformCharacter "A" for each Signal. The waveform behaviors could be made much more clear by using a different WaveformCharacter for some set of waveforms — but then the pattern data would be differentiated as well, and the "single set" of Vector states would need to change on a per-Vector basis as well. That starts down the path toward happier mediums.

9.3 Happier Mediums

Each of the examples above has positive and negative aspects. In the All-Waveforms-in-One-Table example, a distinct reference of each waveform is maintained as part of the Pattern data. On the negative side, without processing the entire data there is no way to determine what Signals actually use what waveforms, or what combinations of waveforms may be in effect at any time. In fact, in this example, there is an inherit expectation that all timing events associated with each Signal need to be supported in the test environment, and short of processing all the Vectors to identify what timing is actually used with what Signal, all the timing does need to be supported.

The One-Waveform-per-Table environment delineates completely every unique combination of timing across all Signals. Both the timing used on each Signal, and the timing across all Signals, is clearly identified by enforcing the use of a single WaveformCharacter. However, the disadvantages of this environment are a reduced ability to interpret the behavior of each Signal in a Vector (since all Signals are represented by the same WaveformCharacter), and a secondary effect known as "timing data explosion". By eliminating the option of a WaveformTable to contain more than one behavior for each Signal, each different combination of events across all Signals requires a complete WaveformTable to represent this new combination. This can cause a combinatorial explosion of timing data. This amount of data is often much more difficult to process by an application.

9.3.1 WaveformCharacter Conventions

There can be a large amount of information portrayed by using multiple WaveformCharacters in the Pattern data. There is no standard or required set of WaveformCharacters to use for particular waveforms, and you must be aware that any set of conventions can be violated in practice, but Table 9.1presents some conventional interpretations on WaveformCharacters. There is nothing that prevents these WaveformCharacters from being used in different ways in any particular test, but different uses may be a cause for confusion without careful review of the Waveform definitions.

TABLE 9.1 Conventional WaveformCharacter Representations

WFC	Conventional Interpretation
0	Drive low; single-event NRZ, may or may not be delayed
1	Drive high; single-event NRZ, may or may not be delayed
Z	Disable the test driver; single-event NRZ, may or may not be delayed
P	Pulse, may be either positive or negative
N	Negative pulse. If used, 'P' is typically reserved for positive pulses
C	Another option for a positive-going pulse ("Clock")
K	A negative-going pulse, when 'C' is used for a positive pulse
H	Measure high
L	Measure low
X	No measure. Primary purpose is to remove any previous measure states.

These conventions present one level of information representation in the Pattern data that can facilitate interpreting Pattern operation. This level of information does not document what is occurring in the device. But it makes the Pattern data both more meaningful and can reduce overall data volume as well.

Another convention to consider is to use uppercase characters for WaveformCharacters that represent high-going device activity, and lowercase characters for WaveformCharacters that represent low-going device activity. This convention was used in the example presented in 9.1.

9.3.2 Common Waveform Groups

There are a set of common collections of the waveforms presented in Table 9.1. These collections are based on common timing between the merged waveforms, and common functionality that generates the waveform in the ATE environment. In particular, the '0' and '1' WaveformCharacters, when defining NRZ waveforms, are often merged together when they have common timing (and many examples in this book use that format). But the 'Z' waveform, even when it is also an NRZ with the same format, is often defined separately. This reflects the fact that the generation of the '0' and '1' waveforms at the ATE uses the same hardware, driving that hardware either low or high. But the 'Z' waveform is generated through different hardware operations in the ATE, and often has different dependencies (and often, different timing as well).

Another commonly-paired set of waveforms is the pulsed clock waveform and the NRZ waveform that is associated with the inactive state of the clock. The pulsed clock WaveformCharacter is not as uniform in application environments, but if we assume for the moment that a 'P' WaveformCharacter represents a positive-going pulse, then this WaveformCharacter is often declared in association with the '0' WaveformCharacter representing the clock's inactive state (as an NRZ-zero waveform).

A WaveformCharacter can only be declared once per Signal per WaveformTable in STIL. The previous paragraphs identified that the 01 and the P0 WaveformCharacters may be merged. But if all three WaveformCharacters are present on one Signal in one WaveformTable, then a decision must be made as to what to associate with what. This issue is discussed in the next section.

9.4 Waveform Grouping/Merging

Consider the waveform definitions:

```
(9.55)      Waveforms {
(9.56)         Some_Sigs {
(9.57)            A { '0ns' U; }
(9.58)            B { '0ns' D; }})
```

As previously identified, these waveforms can be merged to a single waveform definition:

```
(9.59)      Waveforms {
(9.60)         Some_Sigs {
(9.61)            AB { '0ns' U/D; }}}
```

This single waveform definition makes the common activity between these two waveforms more clearly visible. It does not change the functionality of these waveforms in any fashion. This is purely a STIL representational option. It is also not *intended* to represent test resource allocation, however the use of common event times (**`'0ns'`** in this example) between these waveforms may identify in the ATE environment that common timing resources may be applied to these two waveforms.

It is also possible to combine the following waveforms:

```
(9.62)      Waveforms {
(9.63)         Some_Sigs {
(9.64)            A { '0ns' U; }
(9.65)            B { '0ns' D; }
(9.66)            Z { '10ns' Z; }
(9.67)            H { '0ns' Z; '30ns' H; }
(9.68)            L { '0ns' Z; '30ns' L; }})
```

As:

```
(9.69)      Waveforms {
(9.70)         Some_Sigs {
(9.71)            ABZHL {'0ns'U/D/P/Z/Z;
(9.72)            '10ns'U/D/Z/Z/Z;
(9.73)            '30ns'U/D/Z/H/L; }})
```

This is an example of over-merging of waveform definitions. While this is completely proper STIL, it is not the primary intent of the '/' construct to be applied this way. Waveform merging is optimally applied when the event times on each waveform is common between all the waveforms being merged, and also when the event types (at least for the common times) are the same; for instance when the events at a common time are all drive-type events or compare-type events.

Over-merging can be detected when the waveform list has one or more repeated states for an event entry; in this case the first two states (U/D) are repeated across all events in the waveform, and the last two states (Z/Z) are repeated for the first two events. When this situation occurs, the unique characteristics of the individual waveforms are lost to visual inspection behind the redundant events being declared — the test program becomes harder to understand. Also, the timing information may not translate into the ATE environment as directly, as either the translator needs to drop redundant events or the ATE environment needs to support this redundant behavior.

9.5 WGL Pattern Data and WaveformCharacters

WGL does not support user-defined WaveformCharacters, but has a defined, limited set of characters that can appear in the WGL pattern data. These characters have helped to establish the WFC conventions presented in Table 9.1. WGL specifically supports the following characters in the pattern data: '1' to drive or compare high, '0' to drive or compare low, 'Z' to drive off, and 'X' for an undefined waveform.

The WGL waveform definitions support an explicit state-type character to indicate data from the pattern is used to resolve or specify the next operation: an *S* state in a WGL waveform defines a drive substitute operation, and a *Q* state defines a compare substitute. When an *S* or *Q* event is present in the timing defined for a Signal in the WGL context, then the pattern data '1' or '0' is used to indicate the drive high or drive low (for *S*), or compare high or compare low (for *Q*) operation to assert in that waveform for that Vector. STIL does not support this same mechanism of dynamically resolving the waveform activity based on the pattern data, but provides a similar representation of waveform selection and identification via the waveform merging operations presented in this Chapter.

CHAPTER 10 Advanced Timing

Standard
Test Interface
Language

The Timing block contains yet more constructs to facilitate the creation of timing information, and to support additional styles of tests. This chapter will complete the presentation of the remaining Timing constructs and identify their use.

This chapter presents additional Timing constructs. These constructs are part of standard STIL definition, but the use of these constructs are more likely to cause support issues or require interpretation and expansion to support in an application. A robust STIL environment will take advantage of these constructs, but in general the use of these constructs can be considered advanced STIL application.

10.1 Incremental Timing Definitions

We have presented the organization of Timing data around the WaveformTable construct, and the definition of WaveformCharacters inside each WaveformTable for all Signals in a design. Defining timing data for all Signals can result in a lot of information per WaveformTable. Tests that have different timing, expressed in different WaveformTables (with the caveats presented in Chapter 9, Partitioning Timing) needs to define the timing across all the Signals in each WaveformTable. Which creates more timing information to sift through when trying to understand a test program.

Timing data is often redundant, across waveforms and across Signals. A set of Signals with common behavior, for example a bus of signals in a design, tend to share common characteristics, and tend to operate with common timing. Some reduction of magnitude in the timing data can be accomplished by using SignalGroups to reference all Signals that share a common set of timing. Additional reductions can be accomplished by Waveform merging. Each of these reductions has its place — using SignalGroups helps to identify signal associations; merging waveforms identifies common timing elements between waveforms.

One of the requirements identified early in the standards definition was the need for a mechanism to define timing data once and reference it as necessary. The motivation for such a construct came from several of the existing formats, which did not have such a mechanism and all felt the impact of the size of the resulting timing data, and the general expectation that timing complexity in general would be increasing.

The form of this construct in STIL uses the notion of *inheritance*. Inheritance is a behavior of object-oriented languages, where a definition of one object is "inherited" by another object of similar type, and augmented locally. The inheritance concept was seen to be appropriate for this application because of the clearly identified objects and the ability to augment the object, in particular, to define additional definitions or replace definitions previously defined.

The concept of inheritance needs to be applied carefully in this context. While all the information defined in the inherited construct is inherited into the new construct, the replacement or override behavior is limited. Most inheritance behaviors support incremental additions to existing definitions when inheriting a construct. The inheritance behavior in STIL generally overrides the previous definition completely except in one situation, that we will present.

Before we present the Inherit statements we need to identify an additional construct in the Waveforms that hasn't been presented yet. This construct is used to reference waveforms from the Inherit context.

10.1.1 Waveform Labels

The signal reference block inside the Waveform block, as part of the WaveformTable, supports a label as shown:

```
(10.1)Signal_references wave_label { WFCs { events; }}
```

If the context of this statement is not clear, review example (4.2).

This label statement is very different than the labels previously defined in the Patterns section (See "Labels" on page 102). These labels do not have a colon, and the label is found here as the second token after the signal references, as opposed to the Pattern labels that are found as the first token in the statement.

There is another label construct on each event inside a Waveform as well:

```
(10.2)event_label : 'time_expression' state_value(s) ;
```

This label was previously presented at example (8.25) in Chapter 8, Specifications. It is mentioned here for completeness in regards to labels in the Waveforms.

STIL Conventions

The scope of these labels, or the namespace for these labels, is incompletely defined in the STIL standard. The only reference in the Standard is to the event_labels, which are defined to be scoped within the set of Timing referenced in a Spec context (note in this context, these labels need to be unique across the set of Spec variables as well). At the absolute minimum, both of these labels need to be unique across the current WaveformTable where they are defined. As previously mentioned, the event_labels also need to be unique across the set of Timing that may be used in a set of Specs if events are referenced from Spec constructs (see Section 8.9). At the absolute maximum, and to prevent any potential problems with label conflicts, each set of these labels (the wave_labels and the event_labels) may be unique across

```
WaveformTable wft_name {
    ...
    InheritWaveformTable timing.wft ;  ◄──────────────(10.3)
    Waveforms {
        sigref_expr wave_label {
            InheritWaveform timing.wft.wave_label ;  ◄────(10.4)
            WFCs {
             InheritWaveform timing.wft.wave_label.WFCs ;  ◄─(10.5)
             ...
            }
        }
    }
}
```

FIGURE 10.1 Inherit Statement Locations

all Timing blocks present in a STIL test. For the wave_labels, the inherit constructs (defined in Section 10.1.2) support a style of referencing that unambiguously resolves labels down to the WaveformTable level if the labels are not unique across the WaveformTables defined in one named Timing block.

10.1.2 Inherit Statements in the Timing Block

The Inherit statement has three forms depending on where it is found. The statement syntax is presented next, and Figure 10.1 identifies the locations where these statements may be found. The following sections describe the name referencing convention used by these statements and what each statement inherits.

```
(10.3)InheritWaveformTable timing_name.wft_name ;
(10.4)InheritWaveform timing_name.wft_name.wave_label ;
(10.5)InheritWaveform timing_name.wft_name.wav_label.wfc;
```

Each form of the Inherit statement references a different set of timing information; the choice of which statement to use depends on what you want to inherit. We will discuss each of the Inherit operations in turn.

10.1.3 InheritWaveformTable timing_name.wft_name

The InheritWaveformTable statement inherits the contents of an entire WaveformTable into the WaveformTable where this statement is found. The reference to another WaveformTable is resolved as follows:

- The *wft_name* must be specified, and obviously must reference another existing WaveformTable. If the *timing_name* is not specified, then this WaveformTable must be found in this Timing block or in the global (unnamed) Timing block.
- The *timing_name* (and the period) is optional unless the WaveformTable to be found is inside another named Timing block. When present, it references the specific named Timing block where this WaveformTable is found.

Consequently, there are two forms of the inherited reference that is supported by this construct:

```
// Inheriting a WaveformTable from this Timing
// block, or from a global (unnamed) Timing:
InheritWaveformTable "gen_timing";
// Inheriting a WaveformTable from another
// named Timing block:
InheritWaveformTable "all_timing"."gen_timing";
```

The InheritWaveformTable statement inherits the entire contents of the referenced WaveformTable, specifically, the Period statement, the Waveforms section, and the SubWaveforms section (SubWaveforms are described in Section 10.3). Any local definitions inside this WaveformTable will override these inherited definitions, as described in Section 10.1.6.

Perhaps now it is clear why some of these statements may be optional in WaveformTables, because the information may come from Inherited blocks or may be provided for locally and not be defined in an Inherited block.

STIL Conventions

Remember the define-before-use (in this case, define-before-inherit) restrictions. The STIL standard does not state explicitly that WaveformTables that are inherited need to be defined before the WaveformTables that inherit from them, but it is best to generate STIL that enforces this ordering restriction. Circular inherits are easily created (and obviously improper).

By generating STIL with a define-before-inherit restriction, circular inherits can be detected and eliminated.

Be aware of the use of the period ('.') in this context. This operator is also used to concatenate STIL strings together. And it is obvious from example (10.11) that the two constructs around the period are just STIL strings that happen to also be names of STIL blocks. There is an ambiguity in the language with the two applications of this construct in this statement. This ambiguity and resolution strategies is discussed in Section 10.1.8.

The *only* time when just the *wft_name* reference is sufficient is when the inherited WaveformTable is defined in this Timing block or found in the global Timing block. In general the complete reference *timing_name.wft_name* is needed. This form may also be used even when the referenced *timing_name* is the name of the current Timing block; it is not incorrect to reference the current Timing block from an Inherit statement found inside this block, although if there is the potential to rename this Timing block as the STIL test is developed, then these Inherit statements also need to change to reflect the new name of this block.

The InheritWaveformTable statement is the only one of the three inherit statements that can inherit the Period and SubWaveforms definitions (if they are present). If this information is needed in the inheriting WaveformTable then the InheritWaveformTable statement must be used. In particular, if a Waveform is being inherited that contains SubWaveforms definitions defined in the Inherited WaveformTable, then those SubWaveforms definitions need to be inherited as well.

10.1.4 InheritWaveform timing_name.wft_name.wave_label

To use this statement, the inherited information must have a wave_label as discussed in Section 10.1.1. The reference resolution process follows the same procedure as with the InheritWaveformTable statement, with the addition of an additional hierarchical level in the process. To be explicit:

- A *wave_label* is required. If just the wave_label is present then the label must be found in this Timing block (it may be within another WaveformTable) or within the unnamed Timing block if present.

- A *wft_name* may be specified, with a period between the *wft_name* and the *wave_label*. The *wft_name* is needed if the *wave_label* is not unique across this Timing block, or if the *timing_name* is required.
- The *timing_name* is optional but absolutely required if the inherited information is defined in a different named Timing block.

There are 3 forms of the inherit reference here. These forms are absolute, for instance, you cannot skip the *wft_name* if you need to specify the *timing_name*. One field in the statement means the *wave_label* is present, two fields mean the *wft_name* and the *wave_label*, and all three fields means a fully-specified reference. Be aware that the references in this statement reference different objects than the references in the InheritWaveformTable statement, even though they may look the same:

```
(10.12)// Inheriting a set of waveforms:
(10.13)InheritWaveform "basic_inputs";
(10.14)// Inheriting from another WaveformTable:
(10.15)InheritWaveform "all_timing"."basic_inputs";
(10.16)// Inheriting from another timing block:
(10.17)InheritWaveform "C"."all_timing"."basic_inputs";
```

As stated in the previous STIL Conventions section, the reference naming constructs are required if the references need to be unambiguously resolved, but are optional if the names are not required to unambiguously identify the construct.

This statement inherits all the Waveforms, including the WFC definitions, defined in the block identified from the wave_label, and adds them to the definitions for the signal_reference that contained this inherit statement.

10.1.5 InheritWaveform timing_name.wft_name.wave_label.wfc

This statement, at the location it is applied, allows inheriting a single WaveformCharacter or WaveformCharacters definition for a set of Waveforms. Note the WaveformCharacter definitions are not themselves inherited (the statement occurs inside the WFC definitions), just the Waveform definition.

The reference construct for this statement contains another level of naming hierarchy from what is present in the last statement. This statement must contain a reference to a WFC or set of WFCs to isolate a specific Waveform, but follows the same resolution process, as these other processes. So there are four possible forms of reference constructs for this statement:

```
(10.18)// Inheriting from a previous WFC definition:
(10.19)InheritWaveform "H";
(10.20)// Inheriting the WFC in a specified label:
(10.21)InheritWaveform "basic_inputs"."H";
(10.22)// Inheriting from another WaveformTable:
(10.23)InheritWaveform "all_time"."basic_inputs"."H";
(10.24)// Inheriting from another timing block:
(10.25)InheritWaveform "C"."all_time"."basic_inputs"."H";
```

STIL Conventions

The scope of a WaveformCharacter, by itself, is the current set of signals. Typically WFCs are reused between different signals and the reference to another WFC will be ambiguous by itself (unless it is intentionally referencing another WFC in the set of WFCs associated with this signal). This implies that additional reference constructs (at least the wave_label) is commonly required to make useful references outside this context.

A single Waveform definition may support multiple waveforms, using the '/' notation for multiple different event behaviors between the waveforms. When such a Waveform is inherited, the number of WFCs in the definition must match with the number of merged events in the inherited definition.

10.1.6 Inherited Information Resolution

When a block of information is inherited, it is not the block itself that is copied into this WaveformTable, but the *information contained* in that block. So, for instance, when a WaveformTable that contains a Waveforms block is inherited into another WaveformTable that also contains a Waveforms block, there are not two Waveforms blocks in that WaveformTable but the information contained in both blocks is combined to a single set of Waveforms definitions in that WaveformTable.

The primary inherited information is the waveform definitions across individual signals. The inherit statements provide a base set of definitions of WFCs for each signal. The STIL standard defines that inherited information is order-dependent; if the same WFC is defined from multiple inherits, the last defined WFC in a set of inherits will be the one used. Local definitions of the same WFC for the same signal replace any inherited definitions.

When multiple inherit statements are present, the last inherit statement is defined to be:

- The last inherit statement parsed at each of the inherit statement locations.
- The innermost inherit statement. If an inherit statement inside the WFC block is present, that statement causes that waveform for that set of WFCs to be applied.

STIL Conventions

Be careful with the combination of InheritWaveform statements inside the waveform statement, in combination with event statements. at the same location. The STIL standard does not define inherit behavior beyond inheriting waveforms and replacement of inherited waveforms by local definitions; when complete events are defined in the waveform they establish a local waveform definition which replaces, does not augment, the inherited definition.

10.1.7 Inherited signal_reference Resolution

As identified in the previous section, it is the information that gets inherited, and not the syntax constructs themselves. This means that if the inherited information contains signal references (from applying the InheritWaveformTable statement), then these signal references are resolved within the context of the original Timing block, and inherited into another Timing block basically at the Signals levels. An application may or may not maintain the same organization of waveforms (using the same signal reference names) in the Timing block that inherited a WaveformTable, but it can only maintain this information when the signal reference names are identical between the two Timing blocks. If the set of SignalGroups reference statements in each Timing block is different, then it may not be possible to maintain the same signal references in the inherited information.

Even when the SignalGroups reference statements are the same, it is still sometimes necessary to expand a set of waveform references that use a group, into the individual Signals, when data is inherited. If a WaveformTable uses a signal group in a reference to define a set of Waveforms, and inherits information that redefines the WaveformCharacters for a subset of this group, then the original group definition can no longer be applied in this context and the information must be (at least) referenced by individual Signal name after the inherit.

STIL Conventions

Be careful with differences in the resolution of signal references in Timing blocks that will be inherited (with the InheritWaveformTable statement). Inheriting WaveformTables from a Timing block that contains a different set of SignalGroups reference statements may cause confusion when interpreting the inherited information.

10.1.8 The Syntax Issue with Inherit Names and Concatenation

The multiple use of the period character ('.') in the context of inherit statements creates an ambiguous syntax interpretation between separating the levels of hierarchical names and the string concatentation operation. While this situation is ambiguous, there are some simple recommendations to avoid problems here.

First and foremost, there should never be a need to use string concatenation to present any of these hierarchical names. Therefore there should never be a reason to interpret any construct here as string concatentation. This is reiterated in the STIL Convention below.

However, if any name at any particular level does not resolve to the referenced item, then before declaring an error that this item does not exist, it may be useful to consider the potential for string concatenation in this statement. The interpretation of the reference is dependent completely on the number of unique names in the hierarchical reference when this strategy is employed. But it is best to avoid the need for this strategy.

STIL Conventions

Do not use names requiring string concatentation in the inherit statements. Do not define the names being referenced using string concatenation, either. This eliminates the ambiguity issue identified in this section.

10.2 Inherit Strategies

The inherit environment supports a variety of inherit options and behaviors. The following examples present only a partial representation of the capabilities with the inherit constructs. These examples will hopefully open the door to the opportunities of inherit operations.

Inheriting timing information reduces the number of timing constructs present in the inheriting locations, and therefore reduces the overall volume of the data in a STIL test, but only when these definitions are re-used multiple times. It is important to remember that data re-use is a fundamental benefit of the inherit construct (the following examples in this section do not do this in the interest of clarity). Inheriting information also provides a level of "information hiding" from the perspective of a visual inspection of the definitions at the inheriting location, and can make test program analysis more difficult. So the application of the inherit construct has both benefits and disadvantages to consider.

10.2.1 InheritWaveformTable Examples

These examples use the InheritWaveformTable construct because it inherits the broadest set of data from a WaveformTable definition, including all Waveforms, SubWaveforms, and the Period statement — when present.

One capability of the InheritWaveformTable statement is inherit the Period statement. The Period statement is required to be present in any final set of WaveformTables that are used by a PatternExec. But because it can be inherited it may not be physically present in all WaveformTables, relying instead on inherited definitions. Figure 10.2 demonstrates this capability. The converse situation is valid as well; a WaveformTable that contains constructs that are to be inherited by other WaveformTables *only*, may not contain the Period statement itself, relying instead on the Period statement to be present in the WaveformTables that inherited these definitions.

The next example in Figure 10.3 builds on this previous example. This example demonstrates the ability to inherit a set of WaveformCharacters, and in addition to define additional WaveformCharacters on specific Signals in the WaveformTable. In this example, all the WaveformCharacters defined in the WaveformTable "period" in the Timing block "general_timing" are inherited into the WaveformTable "wft_1" in the Timing block "applied", and the WaveformCharacter 'P' is also defined in that WaveformTable, only for the signal "clk1".

```
Timing "general_timing" {
    WaveformTable "period" {
        Period '150ns';
    }
}
Timing "applied" {
    WaveformTable "wft_1" {
        InheritWaveformTable "general_timing"."period";
        Waveforms {
            ...
        }                           Period statement inherited
    }
}
```

FIGURE 10.2 Inheriting the Period Statement

```
SignalGroups {
    "all_in" = '"in1"+"in2"+"clk1"';
    "all_ou" = '"out1"+"out2"';
}
Timing "general_timing" {
    WaveformTable "period" {
        Period '150ns';
        Waveforms {
            "all_in" { 01 { '0ns' D/U; }}
            "all_ou" { LH { '140ns' L/H; }}
        }
    }
}
Timing "applied" {
    WaveformTable "wft_1" {
        InheritWaveformTable "general_timing"."period";
        Waveforms {
            "clk1" { P { '45ns' U; '50ns' D; }}
        }
    }
}                                  "in1" and "in2" support WFCs 0,1
                                   "clk1" supports WFCs 0,1, and P
```

FIGURE 10.3 Inheriting WaveformCharacters

Figure 10.4 extends this example with one more behavior, by demonstrating the ability to incrementally replace inherited WaveformCharacter definitions by local definitions.In this example, the timing defined for the Waveform-Character 'L' for signal "out2" is locally changed in WaveformTable "wft_1" to replace the inherited Waveform definition.

```
SignalGroups {
    "all_in" = '"in1"+"in2"+"clk1"';
    "all_ou" = '"out1"+"out2"';
}
Timing "general_timing" {
    WaveformTable "period" {
        Period '150ns';
        Waveforms {
            "all_in" { 01 { '0ns' D/U; }}
            "all_ou" { LH { '140ns' L/H; }}
        }
    }
}
Timing "applied" {
    WaveformTable "wft_1" {
        InheritWaveformTable "general_timing"."period";
        Waveforms {
            "clk1" { P { '45ns' U; '50ns' D; }}
            "out2" { L { '145ns' L; }}
        }
    }
}
                                          "out1" H WFC event at 140ns
                                          "out1" L WFC event at 140ns
                                          "out2" H WFC event at 140ns
                                          "out2" L WFC event at 145ns
```

FIGURE 10.4 Replacing an Inherited Waveform

10.2.2 Inheriting Separated Time and State in Waveforms

STIL supports an additional inherit behavior that is different than the replacement behavior on waveforms demonstrated in Figure 10.4. This inherit behavior only occurs when the event statements in two waveform definitions (one of them being the inherited waveform) are of a very specific style.

Up to this point, the event statement has always been defined to contain at least two sets of constructs: a time_expression, and an event statement:

```
(10.26)'time_expression' state_value(s) ;
```

In the inherit context, these statements may contain only one of these fields:

```
(10.27)'time_expression' ;
(10.28)state_value(s) ;
```

This style of waveform description is called separate time and shape representation, because the event times are described in one waveform and the "shapes" (sequences of states) are described in another waveform. Some ATE environments maintain waveform descriptions in this context, often where the number of supported shapes is limited but there may be multiple timing options for each shape. An example of this style of inherit is shown in Figure 10.5.

```
SignalGroups {
    "all_in" = '"in1"+"in2"+"clk1"';
    "all_ou" = '"out1"+"out2"';
}
Timing "shapes" {
    WaveformTable "all" {
        Waveforms {
            "all_in" "NRZ" { 01 { D/U; }}
            "all_in" "Clock" { P { U; D; }}
        }
    }
}
Timing "applied" {
    WaveformTable "wft_1" {
        Period '150ns';
        Waveforms {
            "all_in" {
                InheritWaveforms "shapes"."all"."NRZ";
                01 { '5ns'; }}
            "clk1" {
                InheritWaveforms "shapes"."all"."Clock";
                P { '45ns'; '50ns'; }}
        }
    }
}
```

FIGURE 10.5 Inheriting State With Time

When timing and states information is separated, the two sets of information are combined in linear sequence; the first shape statement links with the first time statement, etc. There must be the same number of statements in both waveforms when this style of representation is applied.

One representation of the final waveforms in Figure 10.5 is:

```
(10.29)"all_in" { 01 { '5ns' D/U; }}
(10.30)"clk1" { P '45ns U; '50ns' D; }}
```

10.3 SubWaveforms

There is another option of representing waveform information.This representation supports the creation of short waveforms that are invoked in the complete waveform as needed. These waveforms represent a region of events that are often repeated multiple times to create a complete waveforms. Pulse trains (multiple pulses) may be represented in these constructs through a single pulse definition in a SubWaveform and multiple invocations of this subwaveform in the full waveform.

SubWaveforms require two syntax constructs: the SubWaveform block and the referencing construct to the subwaveform from the waveform context.

The SubWaveform block has the following syntax:

```
(10.31)WaveformTable wft_block_name {
(10.32)  SubWaveforms {
(10.33)     subwaveform_label: Duration time_expression {
(10.34)        event_statements
(10.35)  }
(10.36)}
```

The subwaveform_label is required on every subwaveform declaration, it is not an optional label like labels in other contexts. The Duration identifies the period (or the "subperiod" in this context) of these waveforms. This Duration is used when the subwaveform is referenced iteratively, to identify the period of iteration for this subwaveform. The Duration statement is also required, typically a waveform does not have an event at the end of the waveform and therefore the duration is required to identify the actual end of the waveform.

Only one SubWaveforms block is allowed in a WaveformTable, and when present it must appear before the Waveforms block (as the Waveforms block will reference these definitions).

Only the InheritWaveformTable statement will inherit SubWaveform definitions from one WaveformTable into another. SubWaveform definitions are inherited by the subwaveform_labels; a local subwaveform_label with the same name as one in an inherited block will replace the inherited definition.

SubWaveforms referenced in a WaveformTable must be defined in that WaveformTable; you cannot reference a subwaveform_label from a WaveformTable that does not contain the SubWaveform definition for that label (even if the subwaveform_label is unique across the Timing block).

SubWaveforms are referenced from inside waveforms, as part of the event statements, with the following constructs:

```
(10.37) subwaveform_label ;
(10.38) time_expression subwaveform_label ;
(10.39) time_expression \r subwaveform_label ;
```

Some additional subwaveform constructs are also presented in the next section.

The first subwaveform reference statement (example (10.37)) inserts the events of the referenced subwaveform directly into the referencing waveform, with no changes to those events as they are defined in the subwaveform.

The second subwaveform reference statement (example (10.38)) inserts the events of the referenced subwaveform into the referencing waveform, starting at the time value specified in the time expression. All time events in the subwaveform are incremented by the time value of the start of this subwaveform reference.

The third subwaveform reference statement (example (10.39)) iteratively inserts (repeats) the events of the referenced subwaveform into the referencing waveform, starting at the time value specified in the time expression. All time events in the subwaveform are incremented by the time value of the start of this subwaveform reference *plus* the current iteration count times the duration of this subwaveform. This option is useful for creating trains of pulses, by defining the pulse once and iterating it as necessary in the waveform. The starting time_expression is optional in this statement in which case the iterations start at time-zero in the waveform.

STIL Conventions

The STIL standard does not identify the behavior of multiple SubWaveforms invoked from one waveform context at the same time or in an overlapping time frame. It is expected that this situation is permissible as long as the accumulated set of events across all parallel subwaveforms does not violate the event ordering restrictions identified in Chapter 4, Timing. Alternately it may be recommended to execute only one subwaveform at a time when a sequence of subwaveforms are applied. But there are no STIL constructs to support this behavior, and timing the sequence of subwaveforms is left to the user.

10.4 Multiple Data Waveforms

The last option in the STIL Timing constructs is the option to specify multiple event data per Signal per Vector. This context of data representation allows for a more compact representation of device behavior by collapsing multiple vectors into a single operation, and may also more closely represent the test environment by increasing the flexibility of the options to represent waveforms.

Multiple data representation results in fairly complex waveforms with multiple independently specified events per waveform, and may not be appropriate in all but the most flexible of ATE environments.Direct transforms to this data may be necessary to partition the complex waveform into a series of single-state waveforms, and to partition the Pattern vectors into the corresponding multiple vectors that support these partitioned waveforms.

The multiple data context requires two particular constructs: waveform representation with data selection, and Vectors containing multiple Waveform-Character assignments for signals using these waveforms. We will start with the waveform constructs. Once again, the syntax is defined on the event statement:

```
(10.40)'time_expression' states [ data_select ] ;
```

The addition to this statement is the presence of a bracketed integer value at the end of this statement, after the set of events. This value is used to identify which of a series of WaveformCharacters assigned to a signal reference, will be used to determine the state to be applied at this time.

Multiple data waveforms require merged waveform formatting, with at least two WaveformCharacters defined for the waveform. The selection process through the WaveformCharacters present in the signal assignment in the Vector identifies which of these states to be applied indexed by the data_select value for this event. For example, let us use the following waveform definition:

```
(10.41)"Sig1"{AB{ '0ns' D;'10ns' D/U[0];'20ns' D/U[1]; }}
```

The data selection values start at zero to indicate the first WaveformCharacter in the Vector. The waveform does not have to increment these values for each of the data_select fields. There are no order constraints on these data_select values and a previous values may be applied again at another point in the waveform.

This waveform could then be used in a Vector referencing this signal as:

```
(10.42)V { "Sig1" = \w BA; }
```

Assuming this is an individual Signal, then the two WaveformCharacters assigned to this individual signal reference the single merged waveform defined by the WaveformCharacterList AB. The first WaveformCharacter in (10.42) — WFC 'B' — is applied to all data_select fields in the waveform that have the value zero, and the second WaveformCharacter — WFC 'A' — is applied to all data_select fields in the waveform with the value of one.

This creates a waveform on "Sig1" that starts low at time-zero, and then goes high at 10 nanoseconds because the 'B' WFC in the Vector aligns with the second state value in the state-list for this event statement. The the waveform goes low at 20 nanoseconds because the 'A' WFC assigned to this signal aligns with the first state value in the state-list for this event statement. The set of potential waveforms generated by this waveform statement are shown in Figure 10.6.

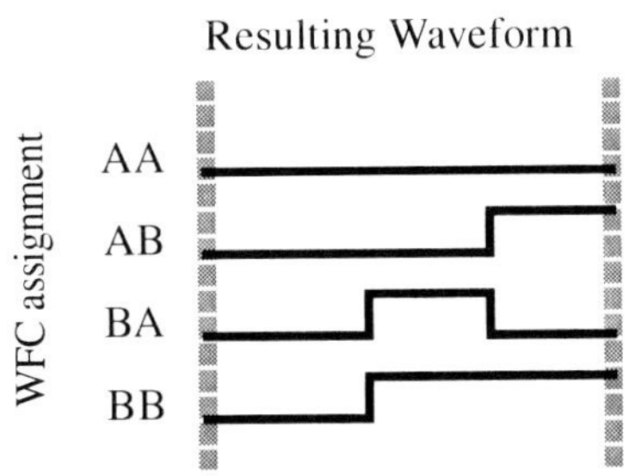

FIGURE 10.6 Waveforms Generated By Example (10.41)

To use multiple data waveforms, all the WaveformCharacters applied to signals in the Vector must all be defined from a single merged WaveformCharacter definition. Also, the data_select construct can only appear in conjunction with multiple-state event statements (event statements containing '/'), and the number of WaveformCharacters in the assignment must match the number of data_select values defined in the waveform.

There is an additional signal assignment construct that may be used in this context as well. Whenever a signalgroup contains multiple Signals, and all are being driven in common with a multiple data waveform, then the following assignment construct is used:

```
(10.43) signalgroup { WFClist_0 ; WFClist_1 ; ... }
```

This is instead of the assignment statement used in (10.42). In fact, that statement can be re-written as:

```
(10.44)V { "Sig1" { B; A; }}
```

Both of these forms are supported for individual signal references. But when signalgroups are present, the form shown in (10.43) is required to unambiguously identify the set of WaveformCharacters associated with each data_select value. Each WFCList statement in (10.43) defines the set of WaveformCharacters to assign to each signal in that group for that data_select value.

This form of multiple data representation is useful in the context where a waveform is seen to consist of segments of time, where each segment may have separate behavior and all segments support the same set of behaviors. And where the total set of behaviors across the stimulus identifies a large number of waveform permutations that may be more cohesively presented with the multiple data constructs.

Finally, the STIL specification identifies how multiple data constructs may also be present in the SubWaveforms. Review the specification for those details.

CHAPTER 11 Procedures and Macros

Standard Test Interface Language

STIL supports a level of structured test development by using procedures and macros. Appropriate use of these constructs has benefit to the overall test program, as we will identify in the next section.

The next section presents a perspective on test development — and how procedures and macros fit into this context.

11.1 Structured Test Development

Test programs are very repetitious. Often the same set of events are repeated over and over again.

The cyclization process takes advantage of this behavior (or requires it to exist, depending on your perspective) at the event level, by collecting a set of repeating events and representing it as a single cycle, and then providing a mechanism to pass specific signal behaviors as collections of those events into a test cycle. But even cyclized data — the Vectors themselves — can generate blocks of repeating operations as the device executes the same behavior over and over again, perhaps with different WaveformCharacters on a subset of the signals each time.

These blocks of cycles may be characterized by a specific device operation, for instance reading off-chip memory, or perhaps the blocks represent a repetitive test operation such as scan data shifting, discussed in detail in the next Chapter. Whatever their existence, STIL supports constructs to represent this data for what it is — a device (or test) function. STIL then allows you to define the functional behavior (as test Vectors) once, and invoke this function whenever the device needs to repeat this same behavior. There are two constructs to define these functions, Procedures and Macros, which will be presented in turn.

The advantage to using procedure and macro constructs are many. First, there is a primary reduction of test data volume, as the repeated sets of Vectors are expressed only once (remember back to the STIL scope: this was one of the goals of the standard). Secondly, test operations are more obvious in the Pattern data when they are expressed by a named reference to these functions, and interpretation (particularly human interpretation) of a structured test program is much more direct. A third and not so obvious benefit is the ability to reuse a Pattern written with structured references, into different contexts or applications of the same Pattern. This third benefit is employed by the Core Test Language (CTL) extensions to support the definition of retargetable Pattern data. Finally, if the test environment supports mechanisms to implement these function calls as well, then applying these constructs can make more efficient use of test capabilities as well.

There is a disadvantage with the presence of procedures and macros in some contexts, but it is easily addressed. If the application environment does not support these function operations for what they are (execution of a single definition), then the functions will need to be expanded into the data before the data can be applied within this application. This is a disadvantage as the application then sees all the "flattened" data. However, expanding these functions is a very straightforward process, slightly more complicated than reading the Pattern data which must already happen (we cover the additional features that make Procedure and Macro processing more complicated, below). The capability of expanding Procedures and Macros is a necessary technique to support incoming STIL containing Procedures and Macros when the application is lacking corresponding support for them.

The following sections use the term "function" to indicate the behavior of either a procedure or macro. This is not the mathematical definition of a function as these constructs do not return a single value back to the calling environment. It is just the term used here to indicate a procedure or macro, interchangeably.

STIL Conventions

If an application or environment does not support Procedures or Macros, or a particular aspect of these constructs, then the Procedures or Macros used in a STIL test, either individually (by individual Procedure or Macro), or collectively (all Procedures or Macros or all that contain a particular construct), may be expanded directly into the Pattern data. This allows Procedures and Macros to appear in any legal STIL context, and still be supported by the application even if the application does not have the capability to support the Procedures or Macros directly. Presence of Procedures and Macros in a STIL test is *not* an issue for test transportation.

11.2 STIL Procedures

The STIL Procedures is a main STIL block, with the following syntax:

```
(11.1)Procedures optional_procedure_block_name {
(11.2)   procedure_name {
(11.3)       pattern_statements
(11.4)   }
(11.5)}
```

Procedure block names follow the naming conventions of most STIL blocks: a single unnamed Procedure block may exist, and if it does all Procedures defined in that unnamed block are accessible from all Patterns. If the Procedure block is named then it must be explicitly referenced to be available in a specific Pattern. Referencing Procedure blocks is discussed in Section 11.6.

The Procedure block contains user named blocks (and all blocks have names; there is no such thing as an unnamed procedure). Each block defines a single procedure, and each block contains any combination of the STIL pattern statements, such as Vectors, Conditions, WaveformTable references, etc. Whatever combination of pattern statements are necessary to implement the desired function are found here. In addition these blocks may also contain the Shift statement that is defined in Chapter 12, STIL Scan.

11.3 MacroDefs

At this level of description, Macros are defined almost identically to Procedures, replacing the Procedures keyword with MacroDefs. Naming conventions on the MacroDefs blocks is the same as for the Procedures, as is the naming of the macros themselves.

```
(11.6)MacroDefs optional_macrodef_block_name {
(11.7)    macro_name {
(11.8)        pattern_statements
(11.9)    }
(11.10)}
```

An important note about STIL Macros: the STIL Macro concept is at the syntactic level of the STIL language, and not the lexical level. This means a Macro function will always contains a set of STIL statements that are complete from the syntactic perspective. This is unlike many computer languages that support the macro as a preprocessed construct and allow macros to be inserted at the lexical level of the language. This same issue was previously identified with the Include statement, and the justification is the same: there is no preprocessing of STIL data necessary to parse the language completely. STIL Macros do not need to be preprocessed into the calling environment because they consist only of complete STIL statements.

So now that we have identified the similarities between Procedures and Macros, we need to identify the differences.

11.4 Differences Between Procedures and Macros

The STIL standard indicates that Procedures are stand-alone entities meant to be "called" from a Pattern context, whereas Macros are meant to be interpreted "in-line" in the Pattern data. The internal operation of the function is the same whether it is implemented as a Procedure or a Macro. What is different between the two is how the function starts from it's calling context (how much information from the calling context is transferred into the function), and what the function returns to the calling environment at the end of execution.

When a Procedure is called, no information about the current environment is transferred into that Procedure. The Procedure starts execution "fresh", which really means that additional information must be provided at the start of the Procedure: a WaveformCharacters for all Signals used in the Procedure must be specified, and a WaveformTable reference must also be defined by the first Vector in the Procedure.

The Macro, by contrast, maintains the state of the calling environment when it starts. Another way to look at this operation is that the Macro maintains the STIL concept of incremental behavior, by maintaining the information of the calling environment. Only *changes* to the information, different WaveformCharacters on Signals or a change to the WaveformTable reference, need to be specified in the Macro.

When a Procedure ends, the flow of the program returns back to the next statement after the Procedure call. The context of the statements following the procedure execution are the same context as the last statement before the Procedure call — all WaveformCharacters on all Signals in the test, and the active WaveformTable reference —are all restored to the values they had before the Procedure was executed, when the Procedure ends. This is not as if the Procedure never executed; the effect of the Procedure persists on the internal elements of the device under test. But the state of the test flow for the next operation is restored on returning from a Procedure. The effects of a Procedure do not persist on the Signals of a design; they effect the internal state only. A Procedure does not return these changes back to the calling environment no matter how the Signals may have changed during the Procedure.

This is again where the Macro behavior differs from the Procedure behavior. Any changes the Macro makes to the Signals in the design, and any WaveformTable reference changes, are all returned back to the calling environment. Once again, the Macro behavior can be seen to be more aligned with the STIL notion of incremental changes, as the changes to the Signals and Timing information during a Macro operation do propagate back into the calling context as if the Macro had been expanded directly into the location where it was invoked.

As a test creator, you have the option to use either Procedures or Macros when you create a STIL test. Which one you choose depends on your requirements on how the function may be defined, and how it may be used. As we stated above, the difference between Procedures and Macros is only on the boundary of the function: how it starts, and what it returns. This difference is fairly subtle, and is best demonstrated by example.

```
MacroDefs "Mac_A" {
    "A" {
        V { "one" = 1; }        Proper Macro Definition
        V { "two" = 1; }
    }
}

                                Incorrect Procedure Definition:
Procedures "proc_A" {     <----   Missing WaveformTable
    "A" {
        V { "one" = 1; }
        V { "two" = 1; }  <----   Improper Signal reference
    }
}
```

FIGURE 11.1 Procedure and Macro Differences

Figure 11.1 presents an identical series of pattern statements as both a Macro body and a Procedure body, and identifies two issues with the Procedure definition presented here. Remember that no state information from the calling context is carried into the Procedure. Therefore the Procedure operates much like a miniature Pattern. Before the first Vector in the Procedure, a WaveformCharacter must be specified on all Signals used in the Procedure, and a WaveformTable reference must be specified. There is no WaveformTable reference in the example above, and there is no WaveformCharacter specified for Signal “two” in the first Vector.

The similarly defined Macro is correct, because the Macro context maintains the environment that it was called from. Therefore the WaveformTable that was last applied in the calling context is propagated into this Macro, and there is no need to specify an initial WaveformCharacter on signal "two" because the last assigned WaveformCharacter is maintained on this Signal.

A corrected Procedure is shown in Figure 11.2.

```
Procedures "proc_A" {
    "A" {
        W "timing_1";
        V { "one" = 1; "two" = 0;}   Proper Procedure Definition
        V { "two" = 1; }
    }
}
```

FIGURE 11.2 The Procedure, Corrected

These behaviors of the Procedure have implication in choosing whether to use Procedures or Macros. Here are some aspects to consider when choosing whether to use a Procedure or a Macro:

- Does the timing in the function change between different applications of this function (for example, is the timing in the function part of the current execution, or is it specified distinctly for this operation)? If the timing environment for this function changes in different applications of this function, then either a Macro should be used (and the timing specified by a WaveformTable reference before the Macro call), or you will need multiple Procedures to support each different timing context for this function.
- Are all the Signals used in the function defined to a set value (or can be set to a defined value) on entry to the function, or does the function operation depend on the state of the signals from the calling environment? This issue can be partially addressed through the use of parameters (presented in Section 11.5), but sometimes parameters can make the operation of a function less clear, or may not be desired in the function. In that case, again, the Macro construct would probably be preferred.

Figure 11.3 highlights another aspect of the different behavior between Macros and Procedures.

```
MacroDefs "Mac_A" {
    "A" {
        W "timing_1";
        C { "one" = 1; }
        V { "two" = 1; }
        W "timing_2";
        C { "two" = 0; }
    }
}
```

Appropriate Macro Definition

```
Procedures "proc_A" {
    "A" {
        W "timing_1";
        C { "one" = 1; "two" = 0; }
        V { "two" = 1; }
        W "timing_2";
        C { "two" = 0; }
    }
}
```

These two statements do not affect test operation because the Procedure does not return state back to the called environment.

FIGURE 11.3 Another Example of Procedure and Macro Differences

The example of Figure 11.3 highlights an aspect of the difference between how Procedures and Macros return information to the calling environment. Macros return their final state (collection of WaveformCharacters on Signals, and the last WaveformTable referenced) back to the calling environment — whether these values have been applied in a Vector not.

The Procedure does not return any state information back to the calling environment, in fact, on returning from a Procedure, the context before the Procedure call (the last assigned WaveformTable and the WaveformCharacters assigned to each Signal) is restored. These values may be changed before the next Vector is executed (of course), but the important thing to remember is the Procedure does not return any state information back to the calling environment.

Which again identifies this single issue to consider when contemplating Procedures or Macros:

- Does the function need to return state back to the calling context, or can the next set of operations be expressed without dependence on the state from this function?

The previous examples highlight the negative aspects of Procedures. So why use Procedures instead of Macros? What are the advantages of Procedures?

Procedures are stand-alone; a well-formed procedure does not depend on the context of the calling environment in order to execute, nor does the environment depend on the final state of the procedure to continue operation. Generally, this makes procedures more portable across different applications as the procedure operation is self-contained. As long as the application environment can support the constructs, procedures are likely to be maintained not only through the flow but all the way through the application. Because the procedure is self-contained, the ability to execute the procedure as a procedure ("calling" a different segment of patterns and "returning" from this section back to the main set of pattern data) is less dependent on how this call/return operation is implemented, and less dependent on the particulars of the test environment.

Macro behavior is less likely to be supported by application contexts, which makes them in one view to be less transportable. This is why the STIL standard identifies that macros are expanded when patterns are processed. Because of the dependency of the of the macro on the calling environment when the macro starts executing, and the dependency of the calling environment on the macro when the macro returns, processing patterns containing macros can be more difficult to maintain by an application. Incremental changes to the Pattern data around Macro calls can affect the Macro behavior.

There is another advantage of using Procedures. Because Procedures are self-contained (at least when properly defined), when the set of Signals used in a Procedure is a subset of the design Signals, the critical aspects of the function operation are more clearly defined in the Procedure. Because a Macro may always depend on Signals not defined in the Macro but set externally before the Macro is called, you can never be sure of the function of the macro — and never be as sure that the macro will function in all environments.

When properly defined, Procedures may be supported in a STIL processing flow as if they were Macros because all information that affects the function performed by this Procedure is contained in that Procedure definition — and in-lining that operation does not change that operation. Macros can not, in general, be implemented as Procedures because the Macro may not be complete by itself. For instance, Macros may rely on the state of other Signals in the execution context to operate properly, and when those other Signals are not provided to this partial function definition, the purpose of the function may not be accomplished.

STIL Conventions

Procedures are intended to be presented in STIL as complete, self-contained entities. It may be possible in some situations, to not specify a particular Signal's WaveformCharacter in the Procedure because of Default-State settings or other mechanisms that provide an appropriate state for that Signal. It is not recommended that a Procedure be structured this way. If the proper operation of a function is dependent on a Signal (and the Waveform-Character or event state associated with that Signal) then that Signal should be part of the Procedure definition. All Signals that affect a Procedure operation should be defined as part of that Procedure to *explicitly* define a self-contained Procedure.

11.5 Procedure and Macro Parameters

Both Procedures and Macros support passing unique Signal WaveformCharacter values into each distinct invocation of the function, to be applied to specific Signals when this function executes. When unique Signal WaveformCharacters are specified with the invocation of the function, each function operation may be different. These unique WaveformCharacters provide a range of different function operations, and the function is said to be parameterized (that is, parameterized around this range of different function operations). The WaveformCharacter values to apply to Signals are called function parameters.

Function operation is sequential through a series of STIL statements. The Signal parameters, to be effective, need to be applied at specific times inside this function. It is not sufficient (normally) to apply all the parameters when the function is invoked, but the proper point to assert each parameter must be identified in the function.

This is provided in STIL through the use of two special assignment characters, '#' and '%'. When a Signal is assigned one of these characters, it means "use the value for this Signal found in the invocation of this function at this point". These two operators are not WaveformCharacters themselves, but they specify the assignment of WaveformCharacters to the referenced set of Signals.

STIL Conventions

Because the '#' and '%' are operators, and not WaveformCharacters, these operators do not persist across Vectors. The '#' and '%' operators do not operate in an incremental fashion, but require an explicit reference to the operator each time a parameter assignment is made. The WaveformCharacter assigned because of this operation is incremental, of course, and does carry forward into subsequent Vectors until changed to another value.

11.5.1 Single Parameter Application

Figure 11.4 provides an example of parameter passing in Procedures. This operation is identical in Macros.

```
Timing {
    WaveformTable "timing_1" { Period 100ns; ...}
    WaveformTable "timing_2" { Period 50ns; ...}
}
Procedures "proc_A" {
    "capt1" {
        W "timing_1";
        V { "all_inputs" = #; "all_outputs" = #; }
    }
    "capt2" {
        W "timing_2";
        V { "all_inputs" = #; "all_outputs" = \r20 X; }
        V { "all_outputs" = #; }
    }
}
```

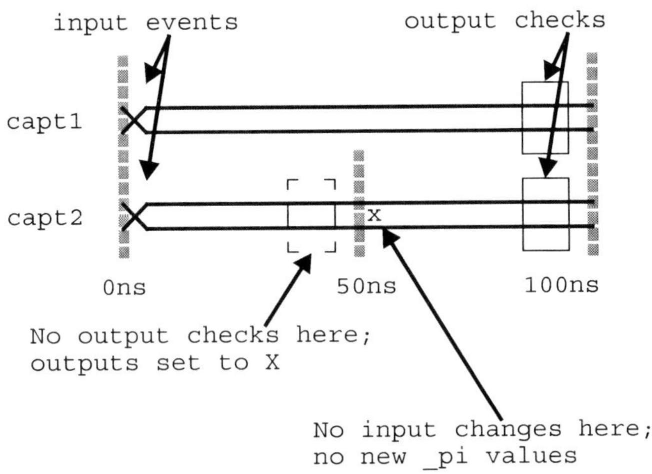

FIGURE 11.4 Single Parameter Application in Functions

Figure 11.4 defines two procedures — "capt1" and "capt2". Each of these procedures operates on the same set of parameters, which are the parameters "all_inputs" and "all_outputs". However, "capt1" passes both parameter values in a single Vector whereas "capt2" uses 2 Vectors to pass this information. Because "all_outputs" is not specified until the second Vector in "capt2", these Signals must also be set in the first Vector, to a value that will not affect the test operation (which is the WaveformCharacter 'X' for these Signals in this example).

Notice that the period of the Vectors used in each of these Procedures is different because two different WaveformTables were used; the period of the single Vector in "capt1" is twice as long as the period of each of the two Vectors in "capt2". So the net execution time of each of these procedures is the same.

Now, if the timing of the events on the Signals in "all_inputs" and "all_outputs" is also similar across these two WaveformTables, as shown in the waveforms in Figure 11.4, then these two procedures are identical in operation. They are not equivalent in capability; "capt2" has an opportunity for additional functionality (at the expense of additional Vectors). But because additional parameters were not specified in "capt2", the behavior between these two procedures is identical.

This is another example of how STIL supports multiple ways of expressing the same behavior. This example highlights in particular a trade-off between Timing (period values) and the number of Vectors. Of course, in this example, setting the period of "timing_1" from 100 ns to 50 ns would allow "capt1" to execute in half the time (and with half the Vectors) of "capt2". But that is not always possible, for example if the outputs need more than 50ns to stabilize before they can be tested. The time between the input events and the output checks must be long enough for the outputs to stabilize, and one technique to allow this is the use of extra Vectors, particularly when the Timing flexibility is constrained.

We've presented how parameter locations are identified in the function body. Passing parameters into STIL Procedures and Macros is covered in Section 11.6.

STIL Conventions

The WaveformCharacter assignment operation in STIL requires a WaveformCharacter to be assigned to each Signal in an expression. This is *not* a requirement when assigning the parameter-reference character '#' and

'%' to Signals or SignalGroups. When assigning a '#' or a '%' to a SignalGroup, a single '#' is sufficient when all Signals are being assigned parameter values. The next two statements (assuming the SignalGroup "_pi" contains 231 Signals) are equivalent:

```
(11.11)"_pi" = #;
(11.12)"_pi" = \r231 #;
```

While it is possible to assign a parameter to a specific Signal inside a SignalGroup, and assign all the other Signals explicit WaveformCharacters, this style of mixing assignments is not recommended. An example of this is:

```
(11.13)V { "_pi" = 0101 # 1110 \r222 1; }
```

This construct is not recommended because it is difficult to identify the Signal that was assigned the parameter value in this expression, which makes review of the test operation more complex. An alternate method to provide this same behavior (and still assign explicit WaveformCharacters with the group "_pi") is:

```
(11.14)C { "_pi" = 0101 0 1110 \r222 1; }
(11.15)V { "SI" = # ; } // WFC '0' in previous C replaced
```

11.5.2 Multiple Parameter Application

A function may need to set a Signal to multiple WaveformCharacter values (in sequence) to accomplish the required operation. There is no issue with assigning the same Signal to different defined WaveformCharacters as necessary.

It is also possible to set the same Signals to parameters multiple times in one function. This is the purpose of the two parameter arguments, '#' and '%'.

The next Figure shows an example of multiple parameter values in the function.

```
MacroDefs "Mac_A" {
    "A" {
        C { "one" = 1; }
        V { "one" = %; "two" = #; }
        V { "one" = 0; "two" = 0; }
        V { "one" = %; "two" = #; }
    }
}
```

FIGURE 11.5 Multiple Parameter References

The difference between '#' and '%' is only detectable on multiple parameter references. A '#' parameter always takes the next value in the passed-in parameters for this reference. That means the parameter may always change on each subsequent '#'. It also means that the function invocation must define more WaveformCharacters to be assigned to this reference, than the size of the reference itself.

The '%' parameter does not change values on multiple references. In fact, this operator causes the same value to be maintained on each operation that uses the '%'. The reason to use this operator is optimization: if a Signal needs to be specified as a parameter, but it holds the same value through several references (sometimes with intervening changes, as when reference "one" is set to 0 in the second Vector of Figure 11.5), then using the '%' parameter permits specifying this value only once in the parameters, and not requiring repeating the same value over and over again in the invocation.

STIL Conventions

Parameter values may be passed to multiple references through either the '#' or '%' operator. That means the same reference name may be used multiple times at different locations in the function body, and each time assigned either the '#' or '%'. It is not recommended to switch between these two argument types on the same Signal. If that Signal needs to be specified multiple times then it should uniformly use either a '#' or a '%' operator for all references.

11.6 Calling Procedures and Macros

Procedures and Macros and invoked from STIL Patterns each with their own pattern statement — the Call and the Macro statement:

```
(11.16)Pattern patname {
(11.17)   Call procedure_name ;
(11.18)   Macro macro_name ;
(11.19)}
```

When parameters are passed into these functions, the block form of this statement is required:

```
(11.20)Pattern patname {
(11.21)   Call procedure_name { signal_assignments }
(11.22)   Macro macro_name { signal_assignments }
(11.23)}
```

The signal assignment statements in these function invocations are identical to those uses in Vectors and Conditions, however the assigned values are not applied to the Signals referenced in the statement until the '#' or '%' parameter is applied to the Signal in the function body. These assignments should be considered "deferred assignments" to be applied at a later time (and as we will see, sometimes not at all). Unlike assignments in both Condition and Vector statements, which take effect when the Condition or Vector statement is processed, these parameter assignments do not take effect until the specified point in the function; the arguments are not assigned until a Vector containing a parameter reference is executed.

There are several mechanisms that STIL supports for aligning the arguments passed into a function invocation, with the parameters themselves in the function body. These different options are presented in the next several sections.

11.6.1 Parameter Passing By Name

The most straightforward parameter alignment process is by using the same signal names in the Macro call that are used in the function body. Figure 11.6 shows the macro of the previous Figure used this way in a simple Pattern example.

```
MacroDefs {
    "A" {
        C { "one" = 1; }
        V { "one" = %; "two" = #; }
        V { "one" = 0; "two" = 0; }
        V { "one" = %; "two" = #; }
    }
}
Patterns "A" {
    W "timing_all";
    V { "one" = 0; "two" = 0; }
    Macro "A" { "one" = 1; "two" = 00; }
    Macro "A" { "one" = 0; "two" = 10; }
}
```

expanded data

```
V { "one"= 0; "two"= 0; }

V { "one"= 1; "two"= 0; }
V { "one"= 0; "two"= 0; }
V { "one"= 1; "two"= 0; }

V { "one"= 0; "two"= 1; }
V { "one"= 0; "two"= 0; }
V { "one"= 0; "two"= 0; }
```

FIGURE 11.6 Resolving Parameter References

This example shows the application of both the '%' and '#' parameters, and the difference in behavior for each parameter. In particular, signal reference "one" which is assigned with the '%' parameter only requires a single WaveformCharacter in the macro call even though this parameter is used twice; that single WaveformCharacter is applied to all occurrences of the '%'.

Signal reference "two" is assigned with the '#' parameter, and therefore each reference to the '#' requires another WaveformCharacter in the macro call. In this example, the '#' is used twice, so a WaveformCharacterList with two WaveformCharacters is required for this signal reference in the macro call to support all applications of data in this Macro.

It is worthwhile to highlight the Loop construct in this context. It is important to realize that each pass through a Loop with a Vector containing a '#' will use another argument value from the function call, as each execution of that Vector in the Loop is distinct. Figure 11.7 demonstrates this behavior.

```
MacroDefs {
    "A" {
        C { "one" = 1; }
        Loop 5 {
            V { "one" = #; }
        }
    }
}
Patterns "A" {
    W "timing_all";
    V { "one" = 0; }
    Macro "A" { "one" = 10111; }
}
```

expanded data

```
V { "one" = 0; }

V { "one" = 1; } // Loop 1
V { "one" = 0; } // Loop 2
V { "one" = 1; } // Loop 3
V { "one" = 1; } // Loop 4
V { "one" = 1; } // Loop 5
```

FIGURE 11.7 Resolving Parameter References Inside a Loop

Parameters may be specified by using either Signal names or SignalGroup names. This capability allows multiple parameters to be specified on the same Signal, separated or grouped by different SignalGroup references that resolve to the same Signal. This concept is presented in Figure 11.8.

```
Signals { "one" In; }
SignalGroups {
    "A" = '"one"';
    "B" = '"one"';
}
MacroDefs {
    "init" {
        C { "one" = 1; }
        V { "A" = %; }
        V { "B" = #; }
        V { "A" = %; }
        V { "B" = #; }
    }
}
Patterns "All" {
    W "timing_all";
    Macro "init" { "A" = 1; "B" = 01; }
}
```

expanded data

```
V { "one" = 1; } // from "A" %
V { "one" = 0; } // from "B" #
V { "one" = 1; } // from "A" %
V { "one" = 1; } // from "B" #
```

FIGURE 11.8 Defining Multiple Arguments on the Same Signal

11.6.2 Parameter Passing By Contents

STIL does not require that the arguments in the function call and the parameters used in the function body match identically. Parameters may also be resolved by identifying the signal references and in both the function call and the function body, and the common signal references between both references identify the argument being passed.

Content matching is often used with SignalGroups containing multiple Signals. Because the resolution process uses the contents of the references (resolving down to the Signals contained in each reference), several parameter references in the invocation may resolve to the same argument in the function body, or vice-versa. This is shown in Figure 11.9, where the name of the argument in the function body does not match any of the names of parameters in the function call in the Patterns, and therefore parameter matching by content is required.

```
Signals { "one" In; "two" In; "tre" In;}
SignalGroups {
    "A" = '"one"+"two"+"tre"';
    "B" = '"one"';
}
MacroDefs {
    "run" {
        V { "A" = %%%; }
    }
}
Patterns "All" {
    W "timing_all";
    V { "A" = 010; }
    Macro "run" { "B" = 1; "tre" = 1; "two" = 1; }
}
```

expanded data

```
V { "one" = 0;"two" = 1;"tre" = 0;}
V { "one" = 1;"two" = 1;"tre" = 1;}
```

FIGURE 11.9 Passing Multiple Parameters into the Same Argument

This mechanism for parameter passing is naturally ambiguous, and if used the following constraints must be followed:

- Parameter resolution by content requires all signal references in all arguments in the function call to be unique. If the same signal is found in two arguments, the argument selection cannot be resolved.
- Likewise, all parameter references in the function body must reference unique signals. Two parameters referencing the same signal or set of signals cannot be distinguished.

Parameter passing resolves argument alignment with the following two-step process:

- First, all references by name are resolved. All arguments in the function body that match a function parameter in the call to that function by name are always associated first.
- Any remaining parameters in the function body that have not yet been resolved are matched with any remaining arguments, and the contents of each are aligned using the Signals that each argument and parameter reference contains, to identify where the data for arguments that do not have a matching (by name) parameter will use the data specified from the remaining arguments.

11.6.3 Extra Parameters

STIL does not identify an error if there are extra, unused arguments present in the function invocation. While not a STIL error, it can be very useful if an application recognizes unused parameters in a function call and generates a message about their existence. This can help during test development to identify when a parameter name has been misspelled, for example.

11.6.4 Missing Parameters

STIL provides a mechanism for resolving under-specified parameters of a function as well.

When an argument is present in a function body, but that argument is not specified (either by name or contents) in the call to that function, or that argument has applied all the WaveformCharacters defined in the parameter of the call (used all the WaveformCharacters up), then the last explicitly defined WaveformCharacter on the Signals in that reference will be applied. Note for Macros, the last defined state may be the state prior to the Macro invocation, but for Procedures, if this situation occurs it is an error if the Procedure did not define an explicit WaveformCharacter on this Signal in the Procedure body before this point.

The last explicitly defined WaveformCharacter before the argument in the function body, may be in a construct that has no useful purpose otherwise. For example, in Figure 11.10, the Condition statement preceding the Vector that references all Signals in the Procedure "capture" would appear to be redundant since all Signals are assigned parameter values in the Vector. But this value does get applied when the Procedure call does not define a parameter for the Signals contained in the argument "outs".

Also be aware that the last explicitly defined state may be different than the last WaveformCharacter specified in the parameter list, if the function invocation did not contain enough values for this complete function operation.

```
Signals { "one" In; "two" In; "tre" In;
          "four" Out; "five" Out; }
SignalGroups {
    "ins" = '"one"+"two"+"tre"';
    "outs" = '"four"+"five"';
}
Procedure {
    "capture" {
        C { "outs" = XX; }
        V { "ins" = ###; "outs" = ##}
    }
}
Patterns "All" {
    W "timing_all";
    V { "ins" = 010; "outs" = LL; }
    Call "capture" { "ins" = 011; }
}
```

This condition statement serves no purpose other than to provide a default value for the "outs" group when they are not specified in the Procedure call.

expanded data

```
V { "ins" = 010; "outs" = LL; }
V { "ins" = 011; "outs" = XX; }
```

FIGURE 11.10 Missing Parameters in a Function Call

STIL Conventions

Only one reference to a Signal can use parameter resolution by content. If a Signal is referenced multiple times in a function (using different argument names for each reference) then all but one of the references must resolve by name. STIL does not define a resolution mechanism if a Signal appears more than once in content-matching arguments or parameters.

Generally provide an explicit value for all Signals associated with argument values passed into a Procedure (and most generally for Macros as well), before the presence of the '#' or '%' in the body of the Procedure. If there are no Vectors before the first '#' or '%' then a Condition statement may be used to specify this value. Exceptions to this rule are supported *only* when a Procedure is **always** called with a value for that argument. This is both a robust coding practice and reduces confusion on Macro operation when figuring out where a value came from during test execution.

11.7 Using Named Procedures / MacroDefs Blocks

The functions defined in unnamed Procedures and MacroDefs blocks can be accessed from any Pattern block. The functions defined in named Procedures and MacroDefs blocks, to be used by a Pattern, need to be referenced from the PatternBurst.

The same Pattern can be reused in different environments by referencing different named Procedures or MacroDefs blocks.

Obviously, the Pattern re-used with different sets of named Procedure and Macro blocks will makes the same set of Procedure and Macro calls in all contexts. If, by referencing different Procedures and MacroDefs blocks, the Pattern references different function bodies, then that the same named procedure or macro may be found in each different named block. Therefore, Procedures and Macros may be defined multiple times, in different named blocks, as long as these blocks are never all referenced at the same time (in a single context).

To define the set of named Procedures and MacroDefs blocks that are used in one context, Procedures and MacroDefs reference statements are used in the PatternBurst. These statements are part of the environment of the PatternBurst:

```
(11.24)MacroDefs macrodef_block_name ;
(11.25)Procedures procedures_block_name ;
```

The complete context for these statements is presented in Section 6.2.3.

These statements may be used multiple times, to reference as many sets of functions from as many different blocks as necessary. But each Procedure or Macro name can be present only once across the set of blocks used.

Because Macros and Procedures are used by separate statements in STIL, the set of names used by each construct is independent of the other construct. This really means that while you can only define the Procedure name "load_unload" once across all the procedures used in a Pattern, you can define both the Procedure "load_unload" and the Macro "load_unload", and these two definitions are independent.

STIL Conventions

Procedures and Macros are defined in separate name spaces in STIL, which allows the same function name to exist as both a Procedure and a Macro in one application environment. However, some environments may apply Procedures and Macros using similar support constructs. This process may be complicated if the Procedures and Macros use the same names. Also, errors in program generation are more difficult to identify if the name is correct but the calling construct references the wrong version of the function. It is recommended that Procedures and Macro names be maintained as single set of distinct names to minimize support issues with the test program.

11.8 WGL Procedures and Macros

WGL supports both procedures and macros constructs, very similar to STIL, but the procedure and macro constructs in WGL are more differentiated from each other. WGL procedures do not support parameters, and they are defined *after* the pattern construct has referenced them. They represent a level of behavior in the test environment, where the procedure is implemented by jumping to a subsequent location in memory and returning from that location at the end of the procedure definition.

WGL macros are true lexical constructs. They may contain partial WGL statements (like a segment of a vector), and require a preprocessor environment to support completely. But WGL macros do support parameter values. However WGL parameters are passed as a single value, are resolved by name only, and are applied completely each time they are referenced in the macro body. The notion of passing multiple values and using an operator (like '#' in STIL) to pick the next value is not supported with WGL macros. WGL macros may exist in a source file, but after processing the WGL file the macros are no longer present in the flow.

CHAPTER 12

STIL Scan

Standard
Test Interface
Language

Scan-based testing is a robust methodology employed by many EDA environments to support automated test pattern generation. STIL has specific constructs defined to address the needs of scan-based design and scan test operations.

The next section introduces scan concepts and defines the terminology used when STIL support is introduced in Section 12.2.

12.1 Scan Design

Scan design is a methodology that supports a direct path of data flow through all sequential elements of a design with minimal intervening logical operations (generally except memories, which have separate and distinct test methodologies). By provided a serial data path and clocking all sequential elements in that path, the sequential elements can be "loaded" or set to a defined state by clocking data in, and the current state of all elements can be "unloaded" by observing the state of the last sequential element in the chain as each of the element states shift through. The basic design constructs to support scan operation are shown in Figure 12.1.

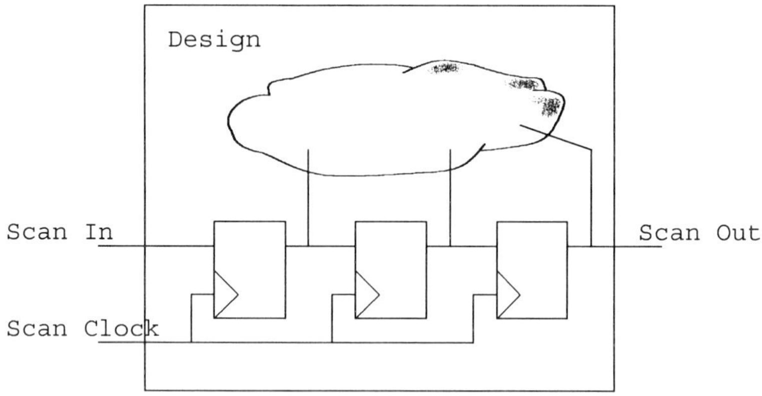

FIGURE 12.1 Scan Design Fundamentals

12.1.1 Scan Operation (Design)

Scan is not the normal mode of device operation. The ordering of all the sequential elements and the direct access through the constructed chain of these elements does not make for a very useful design (other than a very long shift register). A scan mode must be incorporated into the design to support this behavior. Therefore, all sequential elements of the design need to support both modes of operation - normal, functional mode, and scan mode. Which means that the sequential elements have two modes of operation, and are referred to as scan cells to acknowledge the scan aspects and the functional aspects of these cells.

At this point we have established the vocabulary of scan operation: *scan mode* establishes *scan chains* of *scan cells* that are *loaded* through serial *shifting* of *scan data*, and *unloaded* along the same path to observe the states of those elements.

The goal of scan operation is two-fold. First, by serially shifting a defined state into all the scan cells, a high level of design controllability can be accomplished. Second, by serially shifting the current state of the design out of the scan cells, design observability is increased. *Controllability* and *observability* are the watchwords of automatic test program generation. The primary reason to incorporate scan into a design is to leverage ATPG to generate test vectors for the design with measurable test coverage as presented in Chapter 2, STIL Test.

Scan tests are structural tests strategies; they validate that the device is structurally proper and complete.

12.1.2 Scan Process (Test)

ATPG test patterns (and manual patterns, too, I would guess) use the following sequence of operations to perform a test. First, a state of the design is *loaded* with the design in scan mode. Then the design is taken out of scan mode, placing the design into functional or system mode, and one or more "system" clocks are applied to advance the state of the design through the normal design logic paths. This operation is referred to as *capture* as conceptually the system clocks "capture" the expected next-state of the design in all the sequential elements (or the set of elements affected by the system clocks during this capture). Then the design is placed once again in scan mode and the updated design state is *unloaded* and observed. Mismatches with the expected state during the unload indicate some part of the design that did not behave as expected.

Figure 12.2 diagrams this process and identifies the overlay opportunity presented with each set of scan operations.

Scan load uses the input path of the scan chain to establish the state of the design. Scan unload uses the output path of the chain. As the scan chain is just a big shift register, as one set of scan data is unloaded out the output, another set of scan data can be loaded at the same time into the design.

When the scan load of the next pattern is performed at the same time as the scan unload of the previous pattern, the combined scan load_unload operation is termed "merged".

An extension on the ability to completely overlap the load and unload operations, is the ability to do partial merging of these two operations. As long as the unload operation starts first (the current state starts to be detected as the bits fall out of the Scan Out), the load operation may start at any point. Of course, this style of operation generates more overall test vectors, but certain design scenarios require this operation. IEEE Std. 1450-1999 refers to this style of operation as "skewed load-unload".

12.1.3 Characteristics of a Scan Test

The scan load_unload operation has some characteristic attributes. It generally accesses a limited set of the design Signals. Dedicating signals to scan operation is not economical. Sharing signals with scan input and scan output behavior impacts the optimal performance of the signals, and for that reason the number of signals sharing scan behavior are often limited.

Limiting the number of signals with access to scan data in the design causes the average length of the scan chains (which for effective test generation must access a high percentage of the design's sequential elements) to be longer. This is a fundamental trade-off in scan design: the greater number of scan ports available to the design, the shorter the average length of the scan chains. Fewer scan ports means longer scan chains.

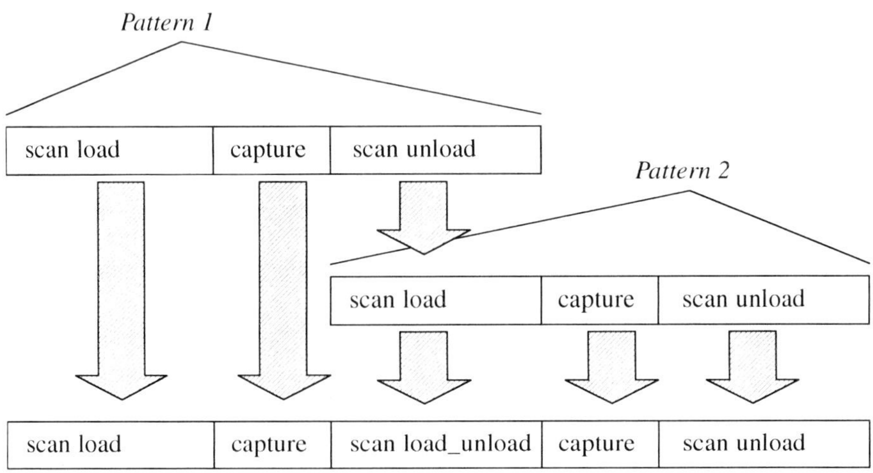

FIGURE 12.2 Scan Test Operation

The length of the scan chains dictates the number of test cycles needed to load and unload the chains, and directly impacts the amount of time needed to perform the scan operation. There are several additional constraints that affect the maximum frequency of scan operation, and consequently the time to perform the scan operation. In many design contexts, the layout of the scan chain is secondary to the primary functional layout, and both the scan data path and scan clock path are not priorities for performance. Also, the scan clock or clocks are often routed to a very high number of sequential elements, all of which are clocked at the same time to create the shift operation. This can have signal integrity and design power consequences as a large number of scan elements change state with the shift. These issues affect the maximum operating frequency of the scan chain, and the amount of time required to complete a scan operation.

Sometimes taking the design into and out of scan mode requires a substantial amount of time as well, as the clock and data paths are re-routed for normal (functional) design operation and the paths stabilize for this mode of operation.

These issues affect the maximum frequency for the scan operation and the amount of time it takes to initiate scan operation or to initiate functional operation for the capture operation between scan loads and unloads.

While scan operation may be constrained on both the number of signals accessed and on the overall performance, the capture operation attempts to leverage as much of the full device behavior as possible. Ideally the capture operation occurs in a context as close to the design operating behavior as possible, or perhaps above the design operating behavior for complete testing — this issue is discussed in the transition testing aspects presented later. And for thorough testing, especially for *all* of the signals in the design, all signals need to be accessed to validate that each is operating properly.

But ultimately it the ATE environment that places the final constraints on the test operation.

12.2 STIL Scan Constructs

STIL defines several constructs to support scan. The primary construct is a special type of loop to support passing scan data from patterns, called a Shift block. The Shift block makes use of two Signal/Group attributes: ScanIn and ScanOut, to aid in manipulating scan data. Then there are some secondary constructs to support scan descriptions that provide additional, optional information on the scan operation.

12.2.1 The Shift Block

The primary construct to support Scan behavior in STIL is the Shift block. The Shift block is allowed inside Procedure or Macro bodies only. This block supports passing scan state data via calls to the Procedure or Macro.

The Shift block operates similarly to a Loop block, except instead of executed a counted number of times, it executes a data-dependent number of times. The greatest amount of scan data on any single argument in the constructs contained in the Shift block dictates the number of iterations for this loop.

An example of a scan-oriented Procedure with a Shift block is shown below:

```
Procedures {
    "load_unload" {
        W "ShiftTiming";
        C { _scanin = 0; _scanout = X; _clk = P; }
        Shift {
            V { _scanin = #; _scanout = #; }
        }
    }
}
```

This procedure is then called in the Patterns, for example:

```
Pattern "sa_tests" {
  C { _all_signals = 001010XXXXH0001; }
  Call "load_unload" { _scanin = 010; }
  V { _ins = 010101; _outs = HHHHXL; }
  Call "load_unload" { _scanin = 011;
```

```
(12.15)        _scanout = HHL; }
(12.16)}
```

Scan data presented to the Shift arguments is sent directly to the Signals referenced in the Shift arguments. The data is presented in left-to-right order; the leftmost WaveformCharacter in the data list is applied to the scan Signals first. In the example above, the first value is checked against the scanout state and the first value on the scanin is applied to the scanin signal.

Since both the scan-in and scan-out data is accessed in the same order for the STIL representation (leftmost WaveformCharacter accessed first), but each of the scan data accesses opposite ends of the scan chain, the scan-in and scan-out data for one chain is ordered the same way when you consider how this data is shifted through the scan chain. Assuming the scan chain is ordered by defining the element adjacent to the scan-in as the first element, and the element adjacent to the scan-out as the last element in the chain, then the first WaveformCharacter in the scan-out data is the state of the last element in the chain because that is the first state on the scan-out, and the first WaveformCharacter in the scan-in data ends up at the last element of the scan chain (because it is placed into the scan chain first, and by the end of the shift operation it ends up at the last scan element).

Timing is everything in a Shift block. The relationship of the scanin assertion timing, the scanout measure timing, and the clock pulse is integral to processing the shift operation properly. The clock pulse causes the next scan data to be shifted out, and the next scan state on the input to be shifted in.

STIL Conventions

STIL supports any Signal or SignalGroup reference in the constructs contained in the Shift block. The reference may contain multiple Signals in the Shift definition. But when the Procedure or Macro that contains this Shift block is called in the Patterns, the signal references in the call should be by individual signal. If multiple signals are referenced in the Shift then the scan data references in the call will rely on content alignment for the signal parameters. To use name alignment of the arguments, individual signal references (Signal names themselves or SignalGroups that are individual signals) should be used.

12.2.2 ScanIn and ScanOut Attributes

A Signal or SignalGroups declaration supports two attributes associated with scan data processing. These are the ScanIn and ScanOut attributes.

Signal references assigned with '#' options inside the Shift block need a ScanIn or ScanOut attribute associated with each one. The signal reference used in the Shift block itself needs the attribute; the constituent Signals in that reference do not need and may not have these attributes.

The ScanIn and ScanOut attributes may specify an optional scan data length as well. This length specifies the amount of data that may be present on the referenced Signals during the Shift.

The purpose of the ScanIn and ScanOut attribute is to establish data padding if the scan data on these Signals is not the length of the longest data in the Shift. Scan data padding is described in Section 12.3.

When these attributes are present on a Signal or SignalGroup definition, the Base attribute may also be used to specify Hex or Dec data formatting and WaveformCharacter interpretation for that Based number.

This is shown in the example on Figure 12.3. The SignalGroups "scanIN" and "scanOU" are used in the Vector in the Shift, and have the necessary ScanIn and ScanOut attributes associated with them. They also have Base attributes defined. When these groups are used in the Pattern, this format is used to represent the data.

```
ScanGroups {
    "scanIN"='"si1"' { ScanIn; Base Hex 01; }
    "scanOU"='"so1"' { ScanOut; Base Hex LHX; }
}
Procedures {
    "load_unload" {
        W "ShiftTiming";
        C { "scanIN" = 00; "scanOU" = XX; _clk = P; }
        Shift {
            V { "scanIN" = ##; "scanOU" = ##; }
        }
    }
}
Pattern "scan_data" {
    C { _all_signals = 001010XXXXH0001; }
    Call "load_unload" { "scanIN"=5; }
    V { _ins = 010101; _outs = HHHHXL; }
    Call "load_unload" { "scanIN"=5; "scanOU"=42; }
}
```

FIGURE 12.3 ScanIn and ScanOut Attributes

12.2.3 ScanStructures Block

The ScanStructures block is an optional block that describes the scan chains of the design. This information is not required to execute a test program, it is not required to interpret the scan operations. It provides additional documentation about how the scan data may be applied to the internal scan cells during the Shift operation.

The ScanStructures block has the following syntax.

```
(12.17) ScanStructures scanstructures_block_name {
(12.18)   ScanChain scanchain_block_name {
(12.19)      ScanLength integer ;
(12.20)      ScanOutLength integer ;
(12.21)      ScanCells list_of_scancell_names ;
(12.22)      ScanIn Signal ;
(12.23)      ScanOut Signal ;
(12.24)      ScanMasterClock Signal_list ;
(12.25)      ScanSlaveClock Signal_list ;
```

```
(12.26)      ScanInversion 0_or_1 ;
(12.27)   }
(12.28)}
```

12.2.4 ScanStructures PatternBurst Statement

When named ScanStructures blocks are defined, these named blocks need to be referenced to be identified as part of the operation of the pattern. These ScanStructures reference statements are placed in the PatternBurst block as part of the PatternBurst context statements, identified in example (6.12) and shown below:

```
(12.29)ScanStructures scanstructures_block_name ;
```

12.2.5 ScanChain Pattern Statement

When ScanStructures are defined in a STIL test, the active scan chain being shifted can be ambiguous, particularly when there are several scan chains that share a scan input or scan output that is used in that Shift. To resolve the ambiguous scan chain, the Pattern statement **ScanChain** is defined before the Shift operation where that chain is applied. The ScanChain statement context was presented in example (5.13) in Chapter 5, Patterns, and is repeated here:

```
(12.30)   ScanChain chain_name ;
```

If ScanChain statements are present in a Pattern block, that reference ScanChains defined in named ScanStructures blocks, then the ScanStructures statement must be present in the PatternBurst(s) that use that Pattern.

The ScanChain statement may help identify what chain is active during a Shift statement. However there are some issues with using this statement, discussed in the next several STIL conventions.

STIL Conventions

The scope of the ScanChain statement — the range of STIL pattern statements over which this statement is effective — is not clearly elaborated in the STIL standard. There is an implicit assumption that statements before the Shift statement affect the next Shift, and that after this Shift the next set of ScanChain statements are associated with the next Shift.

Commonly, multiple scan chains are active in parallel in one Shift operation. The ScanChain statement supports only one ScanChain reference per statement, which means that multiple statements are necessary to represent the set of scan chains active in a Shift. Identifying the collection of scan chains through the use of multiple ScanChain statements, combined with the ambiguity of scope of each statement, further complicates the use of the statement.

The ScanChain statement references scanchain names defined inside ScanStructures blocks. These scanchain names may be common across multiple named ScanStructures block. If this situation occurs, the ScanStructures statement in the PatternBurst is needed to identify which set of ScanStructures are active in that Pattern block, and the set of scan chain names across these blocks need to be unique.

12.3 Scan Data Alignment ("Scan Padding")

When a design has more than one scan chain operating in parallel, and the scan chains are not all the same length (different number of scan elements), then the scan data can be provided or specified in two different ways. One way the scan data can be specified is so that the data is the same length across all the scan chains, even though the chains themselves are not the same length. This format means that additional scan data must be provided for the shorter scan chains in order to match the length of the longest chain being shifted. This additional data is called scan padding data.

It is possible to "pre-pad" or length-align all scan data defined in a STIL test when that test is generated. Motivations for doing this might be to simplify data consumption, and to define a known, deterministic padding operation that is validated as part of the scan operation (pad input values on the shorter chains may be checked by output values present in the padded data).

However, padded scan data in the test program hides pertinent test data for padded scan chains and may complicate interpretation of the data if the interpretation uses structural references.

The other way the data can be provided is to not pad the data, but provide only the amount of data necessary to satisfy the number of scan elements per chain. This reduces the amount of data present in the test program itself. STIL supports this format of data because it defines scan data alignment operations necessary to execute a proper scan shift operation from this format of the data.

Since multiple scan chains are usually operated in parallel, the scan chain with the most number of scan elements (or to follow STIL semantics correctly, actually the signal with the most amount of scan data) dictates the number of scan in and scan out cycles needed to load and unload data across the entire set of scan chains.

The shorter scan chains are therefore padded to fill and empty along with the longest scan chain. The padded data is added to the beginning of the scan-in data loaded, and added to the end of the scan-out data unloaded to compensate for differences in scan chain length. It is added at the beginning of the scan-in data because during the shifting of the longest scan chain, these padded input values will be shifted out and only the actual data to be maintained in the chain will be left. The padding is provided on the end of the scan-out data because the shorter chain will shift all the current states out first, and then the extra padding states will start to be shifted out until the longest chain is completely shifted.

It is hopefully apparent from this discussion that the padding values added to the beginning of the scan-in data on the shorter chains are the same values (albeit with parity changes from any inversions in the scan chain) that are shifted out at the end of the scan-out data.

To make this clear, consider this next example of Patterns calling a load_unload Procedure, where there are two scan chains being shifted in parallel. The scan data is in bold to identify it for this example.

```
(12.31)Pattern "sa_tests" {
(12.32)   C { "_all_signals" = 001010XXXXH0001; }
```

```
(12.33)   load_unload { "_si1" = 0100001010;
(12.34)             "_si2" = 0000000010;}
(12.35)   V { "_ins" = 010101; "_outs" = HHHHXL; }
(12.36)   load_unload { "_si1" = 0110111101;
(12.37)             "_si2" = 0000000011;
(12.38)             "_s01" = HHLHHHLHLH;
(12.39)             "_s02" = HHHXXXXXXX; }
(12.40)}
```

It is not apparent from this data, that the second scan chain is only 3 bits long. So now consider:

```
(12.41)Pattern "sa_tests" {
(12.42)   C { "_all_signals" = 001010XXXXH0001; }
(12.43)   load_unload { "_si1" = 0100001010; "_si2"= 010;}
(12.44)   V { "_ins" = 010101; "_outs" = HHHHXL; }
(12.45)   load_unload { "_si1" = 0110111101; "_si2"= 011;
(12.46)             "_s01" = HHLHHHLHLH; "_s02" = HHH; }
(12.47)}
```

This form of the data does not pad the scan data in the Pattern, but makes use of STIL semantics to resolve the behavior of the shorter scan chain.

Maybe now the intent of the ScanIn and ScanOut attributes is more clear. When scan data is length-aligned for a specific Shift operation, any data added to the scan-out data must be placed at the end of that data (on the right side of the existing data), while scan-in data must be padded from the left, or leading side of the data.

The location to place this padding data can be determined for signals that are of type In or type Out. But signals of type InOut are often associated with scan data processes as well. When these bidirectional signals are used for a scan port, the scan operation will dictate (at least for a specific shift) whether that signal is being used as a scan-in or scan-out, and to place the padding properly for these signals the ScanIn and ScanOut attributes are required.

12.4 More Considerations on Scan Constructs

The STIL support for scan is intentionally broad, to allow for the opportunity to support permutations of the concepts as new variations on the concept of scan are defined. However, these concepts lend themselves to misapplication as well.

12.4.1 Pre-shift and Post-Shift Scan Data Consumption

Scan data may be consumed in a variety of ways inside a Procedure or Macro. If all the scan data lengths across all calls to the function are all the same, and all data is pre-padded or length-aligned, then the Shift operation itself can be implemented as a fixed Loop. Or, the function may consist of a sequence of Vectors that implement the same behavior (this is often called "flattening the scan"). Or, the function may implement a mixture of "flat" vectors and Shift data.

Some contexts of scan operation may be more optimally supported when the "last bit" or the "first bit" of scan data is specified in a Vector outside of the Shift operation. Of course, in STIL, there are several ways this could be accomplished: the separate bit of data could be passed as an independent parameter in the function call, or this bit of data could be passed as part of the scan data, with the expectation that it will be separately pulled out of the scan data when that Vector is executed.

Shift blocks are greedy. They will consume all bits of data passed to them. For that reason there are special semantics on STIL data to establish how many bits of data passed in a function call, continue to be passed into the Shift block. These semantics are discussed in the next section.

But Vectors that consume scan data outside of the Shift block are not unusual, and the following STIL conventions outline some recommendations in this context.

STIL Conventions

If scan data is consumed in pre-Shift Vectors, it is strongly recommended that the pre-Shift Vectors consume scan output data only. Because scan-out data is post-padded if data alignment is needed, consuming output data in pre-Shift Vectors results in a skewed load-unload operation as discussed previously. If scan input data is consumed in pre-Vectors, any padding that may occur in the Shift operation will result in padded data inserted between these pre-Shift states and the data loaded during the Shift, and that is probably not the desired behavior.

If scan data is consumed in post-Shift Vectors, it is strongly recommended that the post-Shift Vectors consume scan input data only. Because scan-in data is pre-padded if data alignment is needed, this again results in a skewed load-unload operation that is not affected if the Shift operation results in the addition of padding data to this chain.

12.4.2 The STIL Shift Calculation

STIL Shift blocks are greedy, and will consume all scan data bits passed into them. However there are requirements in some environments, to change scan behavior on the last scan bit as it gets shifted into the design. Note this is only a problem for bits *after* the Shift operation, because *before* the Shift operation the data bits would have been consumed and never passed to the Shift (using standard STIL parameter operation as previously presented).

The easiest solution to this problem is to not pass the last scan bit as part of the parameters associated with the Shift block, but pass the data as a separate parameter. The issue with this solution is that the context of this data is no longer apparent — the last bit of scan data is not associated with the other bits.

The STIL Shift block already requires processing scan data for padding issues; if the data is not the same length across all Shift arguments then the data needs to be aligned, and it needs to be aligned differently depending on the scan attributes associated with each shift argument.

So the STIL environment defines explicitly the process to determine how much data gets processed by the Shift block, and the calculation takes into consideration not only bits that may be consumed before the Shift block, but bits that are specified on Shift arguments that are consumed in Vectors after the Shift block as well.

This calculation requires identifying all locations where scan data (that is, data that is associated with arguments in the Shift block) is consumed outside of the Shift itself. Normally this can only happen with name-aligned parameter resolution (and it is strongly recommended to maintain name-alignment here). Data is then reserved outside of the data sent to the Shift for application in these contexts. If the data is consumed before the Shift then leading scan data is reserved; if the data is consumed after the Shift then trailing scan data is reserved.

Only the data that remains after this is actually passed to the STIL Shift context. This data will be length-aligned across all the sets of data present for this Shift, using the padding alignment process previously described.

This process is very different than the fundamental STIL parameter processing operation (which consumes the data in linear fashion), but this mechanism is necessary to support some of the characteristics of the scan operation itself. In particular, the definitions in the STIL standard support an environment that can 'unroll' the scan load and unload operations as necessary to support device operation in scan or test debug and analysis as well.

12.4.3 The STIL Scan Pad State

As presented in Section 11.6.4, STIL defines the last WaveformCharacter assigned to a signal before that signal has a '#' or '%' assigned, to establish the default WaveformCharacter value if that signal does not have a parameter value specified in the call.

Padding scan data falls under this behavior. The Shift operation consumes the longest set of scan data available and the signals with less data present apply this definition to establish the state to present during the Shift. However, the STIL Shift calculation does not place this extra data on the end of the sequence of Vectors for all the signals but is dependent on the scan attributes to determine where to place these states.

Scan shift operations on shorter chains can always be validated with a knowledge of the parity of the overall scan chain, as shown in the Procedure example below:

```
load_unload {
  W "ShiftTiming";
  C { _scanins = 00; _scanouts = HL; } //Pad States
  Shift {
     V { _scanins = ##; _scanouts = ##; _clk = P; }
  }
}
```

In this example, the scan input padding state is "0" for these two scan inputs, and the scan output padding state is "H" for the first scan output and "L" for the second. During the Shift operation, any scan-input values that were pre-padded during the Shift, will "fall out" on the scan outputs and be detected by the padded scan-out values. In this example, since the input padded values on both chains is "0", but the outputs have different states, there is an odd number of inversions in one of the scan chains (assuming conventional definitions for these WaveformCharacters).

STIL Conventions

Padding scan data as part of the Patterns ("pre-padding") is not recommended. It increases the size of the STIL test, does not eliminate the need for the consumer to validate and support unaligned scan data, and hides pertinent test data amongst the padded values.

12.5 Scan Groups

Sometimes, a design cannot shift all scan chains in parallel. Sometimes this is caused by design constraints, sometimes by intent. When this happens, and when the scan chains are individually controllable, the test environment may require multiple sequential Shift blocks to shift the sets of scan chains in turn. Each of the collection of scan chains that operate in parallel is considered a scan group.

Shifted scan groups separately is a fairly expensive option in terms of test time, as it is serializing some aspect of the Shift behavior.

When scan groups are present, it is important to remember that each STIL Shift statement is greedy. Each separate Shift block in this context requires a separate set of scan arguments to be present. As long as this constraint is followed, a STIL Procedure or Macro may contain as many serial Shift blocks as necessary.

The next section presents an example of a scan-based test context. No new syntax is discussed here.

12.6 An Example: Transition Testing

To present some of the considerations for an effective test program and the STIL constructs around that environment, a simple example using Figure 2.3, "Example Transition Delay Fault," on page 22, will be expanded. Transition delay was introduced in Section 2.2.7.

The circuit of Figure 2.3 is expanded to include scan flip-flops at each end of the path. This is shown in Figure 12.4. These scan flops are defined to be clocked from separate clocks in system mode, labeled "H" and "I" in this Figure.

To execute this test, all necessary conditions must first be established. These conditions are shown as state values on the other logic inputs. These logic values are established by loading scan data into other scannable elements in the design. Most importantly, the current state of flop-flop "FF1" must be zero, but the input to "FF1" must be one. This condition is necessary to launch the transition event across this logic, and be able to detect the resistive fault as this transition event propagates through the logic.

Once the conditions have been established, the test is ready to be performed. This is done by clocking "FF1", which "launches" the transition event through the circuitry.

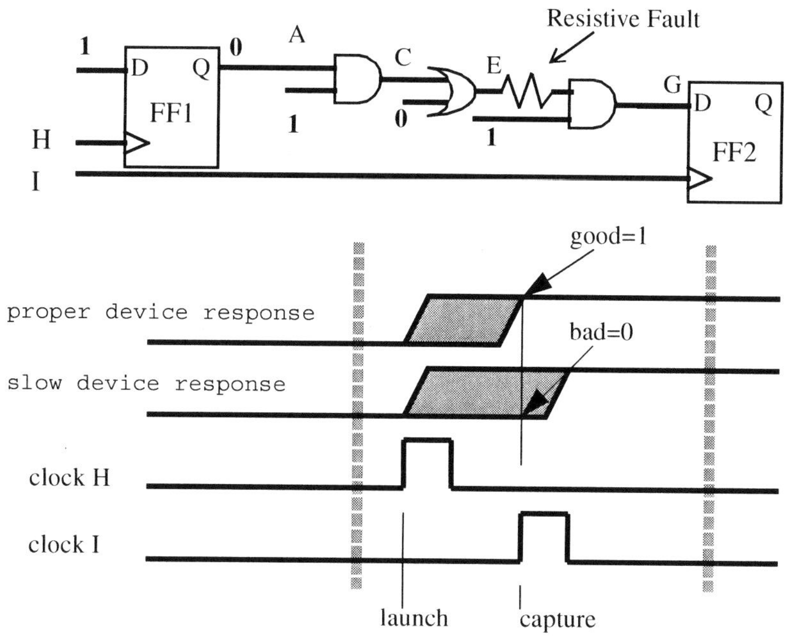

FIGURE 12.4 Example Transition Delay Test

This event must then be "captured" by clocking "FF2" at just the right time offset from the clock on "FF1". If "FF2" captures a logic one at this time, there is no significant delay fault detected from this transition. If "FF2" captures a logic zero at this time, then the path is slow.

Finally, the scan operation must be executed again to unload the state of "FF2" and determine if this fault was detected.

The following STIL code shows one representation of this style of testing. This contains only the main Pattern elements for this test.

```
Pattern "trans" {
    C { "all_sigs"=0000XXX; }
    load_unload { "scan_in"=010...; } // establish design state
    V { "clock_H"=P; }                 // launch operation
    V { "clock_I"=P; "clock_H"=0; }    // capture_operation
    C { "clock_I"=0; }
```

```
        load_unload { "scan_out"=HHL...; }// identify if a failure
        ...
    }
```

Note in this example the explicit maintenance of the clock states: when the capture operation starts, the launch clock is explicitly disabled, and before the next call to load_unload the capture clock is also reset to it's inactive state. Remember that STIL is incremental at the waveform level, and it is important to consider what might happen in the next operation because of Waveforms left on signals in the previous operation.

12.7 WGL Scan

WGL scan constructs are very different than STIL. The fundamental difference between STIL and WGL in this area is with the orientation of the scan data itself. STIL scan data is presented from the perspective of the design signals, which means the scan data is presented ready to be applied in the test environment with no transforms required. WGL scan data is presented from the internal state of the design; the scan data is presented from the perspective of the state of the internal scan cells that hold that data. If there are inversions in the scan chain, then this internal data needs to be transformed before it can be applied in the test environment.

Another difference is in the constructs used to represent scan data. WGL data is represented through an additional level of indirection from the STIL data. This allows the representation of scan data to be more flexible in the WGL environment, but this flexibility is rarely applied because most test generators reestablish the entire scan data on a chain when defining patterns. And while the internal data representation is more flexible, when a scan statement is referenced in a WGL test, it requires identification of a single structural representation of the scan chain to determine how many shift operations to perform. WGL scan data is fundamentally tied to a structural chain definition rather than being dependent on the amount of scan data present, as in STIL.

The WGL *scan* statement itself is different than the STIL Shift block as well. The STIL Shift block supports a variety of test constructs during a Shift operation, in particular more than one Vector can be executed per Shift in STIL. But the WGL scan statement is an extension of the WGL vector statement, and is by definition limited to a single vector of repeated operation.

CHAPTER 13

STIL Levels

Standard
Test Interface
Language

Levels is the STIL terminology for specifying voltage and current values as part of the test information. Levels information is not specified in IEEE Std. 1450-1999, but is defined in IEEE Std. 1450.2-2002, citation [5] .

STIL data without this extension relies on implicit assumptions about the voltage and current values that will be applied to the DUT. This was identified in Section 4.2.1. While this information must be defined before the test is usable, the original STIL development project identified this information as not a substantial part of the test transportation issue. Many test generation flows also do not specify levels from simulation or automatic test pattern generation flows, working with the same assumptions that the levels will be set before the test can be applied.

13.1 STIL Statement with Extensions

Although there have been several examples of the block form of the STIL statement, up to now, the STIL Statement at the top of each STIL file has only required the form:

```
(13.1)STIL 1.0;
```

This identifies a STIL test program containing solely IEEE Std. 1450-1999 constructs.

When the contents of a STIL program include definitions found in STIL Extensions, the STIL Statement is defined as a block statement and includes identifying statements to reference each of the extensions present in this test. In order to define voltage and current information as part of the test program, this requires a statement to reference the IEEE Std. 1450.2-2002 extensions.

To identify the presence of levels constructs in a STIL test, the STIL statement has the form:

```
(13.2)STIL 1.0 {
(13.3)   Levels 2002;
(13.4)}
```

STIL Conventions

It is allowed for a STIL file to identify extensions in the STIL statement, and then never contain any constructs in the file that are defined in that extension. However, it is appropriate for an application to indicate that it cannot support this STIL test and refuse to process this data if the application cannot support the referenced extensions in the STIL statement. This may impact portability of the STIL test even though the test does not make use of these extensions. Identify a STIL test as containing these extensions only when it does contain these extensions.

13.2 Static and Dynamic Levels

STIL Levels defines three distinct capabilities for establishing levels values. The first type of levels is a static behavior, established for each PatternExec execution. Static DC parameters are established with the DCLevels block.

"Static" in this context means "a single value established for the duration of the PatternExec". One set of DC parameters are applied as appropriate, and these values do not change during the test. If a DC parameter needs to change during test operation, then the dynamic levels options must be used.

The second type of Levels capability is a dynamic behavior that supports changing Levels values on Vector boundaries, in the same way that WaveformTables change timing between Vectors.

The last option of levels is an additional dynamic behavior that supports changing Levels inside a single Vector. This requires new constructs inside the Timing Waveforms constructs to define the levels changes during the cycle.

13.3 Static Levels

The primary block for defining DC Levels information is `DCLevels`.This block contains a set of DC parameters, each statement containing one or more dc_expressions.

Dc_expressions are defined similarly to time_expressions, except that they resolve to expression types appropriate for the context of the application — voltage, current, resistance, slew rate (voltage over time), etc. The expressions may contain constants or references to Spec variables. Spec variable support was presented in Chapter 8, Specifications.

The DCLevels block has consistent naming behavior with other STIL blocks. A single unnamed block may exist, and named DCLevels blocks may also exist, which must be referenced explicitly to be applied. Referencing DCLevels blocks for static behavior is defined in Section 13.3.3.

The DC parameters specified in a DCLevels block are applied to individual signals or signal groups. Resolving signal and signal group references in the DCLevels block is performed the same way it is in the Timing block, discussed in Section 4.10.1. As with the Timing resolution and the PatternBurst resolution, signal resolution in each DCLevels section is independent of any other resolutions.

The syntax of the DCLevels block is:

```
(13.5)DCLevels optional_dclevels_block_name {
(13.6)    InheritDCLevels dclevels_block_name ;
(13.7)    SignalGroups signalgroup_block_name ;
(13.8)    signal_expr {
(13.9)        dclevels_statements ;
(13.10)   }
(13.11)}
```

InheritDCLevels is covered in Section 13.7.

13.3.1 DCLevels Statements

DCLevels statements that set parameters have the general form:

```
(13.12)DCLevels_keyword dc_expression ;
```

There is one circumstance where multiple dc_expressions may be found after some of these statements. This is to support intra-cycle levels changing and is covered in Section 13.5. DCLevels statements with dc_expressions are listed in Table 13.1.

There are four statements that consist of a DCLevels_keyword only. These statements set certain behaviors of operation, and are listed separately in Table 13.2.

TABLE 13.1 DCLevels Statements

Type	Keyword	Behavior
Input Voltages	VIH	input high voltage
	VIL	input low voltage
	VICM	common-mode differential voltage
	VID	differential input voltage
	VIHD	differential input high voltage
	VILD	differential input low voltage

TABLE 13.1 DCLevels Statements

Type	Keyword	Behavior
Output Voltages	`VOH`	output high voltage measured
	`VOL`	output low voltage measured
	`VOCM`	common-mode differential voltage
	`VOD`	differential output voltage
	`VOHD`	differential output high voltage
	`VOLD`	differential output low voltage
Slew Rates	`VIHSlew`	input voltage high slew rate
	`VILSlew`	input voltage low slew rate
Output Currents	`IOH`	output high load current
	`IOL`	output low load current
Other statements	`LoadVRef`	load threshold or commutating voltage
	`ClampHi, ClampLo`	voltages for high and low clamps
	`ResistiveTermination`	terminating resistance
	`TermVRef`	terminating voltage
	`VForce, IForce`	forcing voltage, current
	`IClamp, VClamp`	maximum (clamp) current, voltage

TABLE 13.2 DCLevels Override Statements

Type	Keyword	Behavior
Override statements	`ForceHi, ForceLo`	Overrides all pattern data on the Signals; holds Signals at a high or low logic level.
	`InitHi, InitLo`	Holds Signals at a high or low logic level before the PatternExec starts.

The override statements in Table 13.2 are used to control certain test environment behaviors. The first two statements (ForceHi, ForceLo) are used to override the entire pattern data for the signal or signals referenced for this block.These statements are not intended to be applied in a normal test scenario but provide debug control of these signals from the DCLevels context.

The InitHi and InitLo statements specify the state to hold these signals outside of the PatternExec execution.

13.3.2 States and Levels

The STIL standard explicitly correlates STIL states from Table 4.1 with DCLevels constructs from Table 13.1, as identified in Table 13.3.

TABLE 13.3 1450 States and 1450.2 Levels

State	Level
`D` or `ForceDown`	`VIL`
`U` or `ForceUp`	`VIH`
`L` or `CompareLow` `l` or `CompareLowWindow`	`VOL`
`H` or `CompareHigh` `h` or `CompareHighWindow`	`VOH`

Many more STIL states also have relationships with DCLevels statements, but they are not explicitly identified in the standard. The relationships above are tied into intra-cycle level support which is presented in Section 13.5, which is why these relationships are defined.

13.3.3 Applying DCLevels in the Test

By default, DCLevels blocks are not applied to pattern operation. They must be referenced in each PatternExec or in the Pattern, to be applied. *However be aware that an unnamed DCSets block will be applied if this block exists — see Section 13.4.*

To apply a DCLevels block in a PatternExec, the DCLevels block must define all necessary levels statements for all Signals in the design; a DCLevels block that does not define levels for all Signals necessary in a PatternExec cannot be referenced. DCLevels blocks *individually* may define a subset of definitions across a subset of Signals. These blocks may be collected together into one DCLevels block that defines all necessary values across all Signals in the design, using the InheritDCLevels statement.

To specify DCLevels applied as a static level, the DCLevels statement must be specified in the PatternExec. The statement in the PatternExec takes one of the two forms:

```
(13.13)DCLevels ;
(13.14)DCLevels dclevels_block_name ;
```

The first form (without a dclevels_block_name) is used to apply the unnamed DCLevels block to a PatternExec; the second form applies a named DCLevels block to the PatternExec.

This statement can be present only once in a PatternExec; only one DCLevels is applied to the PatternExec in a 'static' capacity.

STIL Conventions

Be careful with your intentions for an unnamed DCLevels block. This block is not applied to a PatternExec by default because of conflict with the default application of the unnamed DCSets block. It is not applied by default in any other context either, but needs to be explicitly referenced. It is recommended to always name the DCLevels block to avoid any assumptions about default unnamed handling.

13.4 Inter-cycle Dynamic Levels

The next level of DCLevels flexibility is the option to specify multiple sets of DCLevels parameters, and the ability to change the current DCLevels active during pattern execution, between Vectors. This flexibility is provided by specifying a collection of DCLevels blocks that may be used together, and a pattern statement to specify the currently active DCLevels block.

There are three distinct constructs to support this dynamic capability. The first construct is the DCSets block, which is used to identify the set of DCLevels that are used in one PatternExec context. The syntax of the DCSets is pretty straightforward:

```
(13.15)DCSets optional_dcsets_block_name {
(13.16)   DCLevels dclevels_block_name ;
(13.17)}
```

The naming of the DCSets blocks follows the normal STIL block naming conventions. There may be one unnamed DCSets block, and any number of uniquely-named DCSets blocks (uniquely-named across the collection of all DCSets blocks). Named DCSets blocks must be referenced to be applied; the unnamed DCSets block is applied to any PatternExec that does not otherwise specify a DCSets block.

There are expected to be multiple DCLevels reference statements in one DCSets block. The DCSets block collects all these DCLevels and identifies them as the set of DCLevels that may be changed during one PatternExec. This serves much the same purpose as the Timing block for Timing information, which collects all the WaveformTables that may be used in one PatternExec, into one location. Except that the DCSets block does not contain the DCLevels definitions, but only a reference to previously defined DCLevels.

The DCLevels reference statement here may also not identify a dclevels_block_name, in which case the statement references the unnamed DCLevels block.

There may be multiple DCSets blocks defined. To apply a named DCSets blocks into a PatternExec, the DCSets reference statement must be present in the PatternExec:

```
(13.18)DCSets dcsets_block_name ;
```

This statement can be present in the PatternExec only once; there can only be one set of DCLevels, identified by a single DCSets reference, active in a PatternExec.

Once the set of DCLevels has been identified in the PatternExec, the final statement controls which one of the DCLevels is active during Pattern execution. This is a pattern statement, similar to the WaveformTable statement in the pattern that specifies the current timing for the next set of Vectors.

The Pattern statement that controls the current active DCLevels block during pattern operation is, once again, the DCLevels reference statement, but this time inside the Patterns block:

```
(13.19)DCLevels ;
(13.20)DCLevels dclevels_block_name ;
```

The first form (without a dclevels_block_name) is used to apply the unnamed DCLevels block to the next set of Vectors; the second form applies a named DCLevels block to the PatternExec.

Remember the previous STIL convention (in Section 13.3.3) cautioning the use of an unnamed DCLevels block. To avoid ambiguities when referencing DCLevels from the Patterns, always use DCLevels that have been included in the DCSets block used for this PatternExec. And provide a name on all DCLevels blocks since there is no benefit from supporting an unnamed DCLevels block, which must also be referenced.

13.4.1 Interaction of DCLevels and DCSets in the PatternExec

There is a fairly subtle interaction of the DCLevels and DCSets statements in the PatternExec that affects Pattern contents, when both statements are present or just the DCSets is specified.

When a DCLevels statement is present in the PatternExec, then the levels specified in that DCLevels block will be applied to each Pattern before each Pattern starts to execute. These levels will continue to be applied until the first DCLevels statement is executed inside that Pattern. In this situation, the DCLevels statement in the PatternExec provides a default set of levels to be applied until another set of DCLevels is specified. This also allows the Patterns to start operation without an explicit DCLevels statement before the first Vector of each Pattern. In fact, it allows a Pattern set that contains no DCLevels statements to operate in conjunction with a Pattern that defines DCLevels internally, with no changes to the Pattern with no DCLevels references — as long as that Pattern can operate from the levels specified in the single DCLevels reference from the PatternExec.

If a DCSets statement is present in the PatternExec without a DCLevels statement, then each and every Pattern applied in that PatternExec must define a DCLevels statement in the Pattern before the first Vector. There is no default DCLevels available to start Pattern execution in this situation. This issue is exactly the same as the requirement to reference a WaveformTable before the first Vector (but with WaveformTables there is no default WaveformTable option, therefore the WaveformTable must always be specified before the first Vector).

13.5 Intra-cycle Dynamic Levels

Some test environments require dynamically changing the DC parameters during a single Vector. STIL supports this capability for a very limited set of DC parameters, using additional constructs in the waveform to reference these DC parameters.

To dynamically change DC parameters during a Vector, multiple DC parameters need to be specified in the current DCLevels block. The VIH, VIL, VOH, and VOL specifications are the only DC parameters that support multiple dc_expressions in the DCLevels statement:

```
(13.21)(VIH VIL VOH VOL) dc_expression dc_expression... ;
```

When this DCLevels block is active in the Patterns (whether specified from a DCLevels statement in the PatternExec, or a DCLevels statement in the Patterns if a DCSets statement is present in the PatternExec), then each of these expressions defines a voltage level that may be selected from within a waveform.

The waveform construct must be extended to support identification of which dc_expression in the list above, to apply to each event as well. This extension is made to the event statement itself, to associate a DC expression with each change of state in the waveform. This is provided by putting an integer number after each STIL state value in the event:

```
(13.22)'time_expression' state_value integer ;
```

The integer value is associated with the state_value, so if multiple state_values are present then each must have an integer value present if any one of them uses one:

```
(13.23)'time' state_value integer / state_value integer ;
```

Only the states listed in Table 13.3 (U,D,H,h,L,l) support an integer value, because only these states identify the DCLevels statement associated with each of these states.

The integer value refers directly to the list of dc_expressions defined for that DCLevels statement associated with the specific state. The value in the list is "zero-indexed", that is, it starts from zero to reference the first element of the list, 1 references the second element of the list, etc.

If an integer value is not specified with a state_value, then the first dc_expression for that DCLevels statement is used (which would be referenced with the integer '0' if it was present).

13.6 DCSequence

The DCLevels standard provides for a construct to address power sequencing issues at test. This is the DCSequence block.

The DCSequence block provides information outside the operation of the PatternExec; it specifies power-up and power-down sequences, and how to handle transitions between different DCLevels environments. As such, this block needs to be applied in the test context but "fills-in" around the PatternExec operation — what to do before, after, and in-between PatternExecs. Therefore this block is applied in a test context along with the sequence of PatternExecs that define the complete device test. That process is considered to be outside the scope of the primary transportation of test data, and therefore what to do with the DCSequence is not defined in the standard.

We provide the framework of the syntax for the DCSequence here, but will not elaborate on this section at this time because of the external connotations with this data:

```
(13.24)DCSequence dcsequence_type {
(13.25)   SignalGroups signalgroup_block_name ;
(13.26)   expr sigref_expr {
(13.27)      Connect connect_type ;
(13.28)      Disconnect disconnect_type ;
(13.29)      Apply dc_expression ;
(13.30)      Ramp time_expression dc_expression ; } }
```

The name of this block is one of a set of defined dcsequence_types. This identifies that each of these types can be defined only once in a STIL test. The defined *dcsequence_types* are: **InitialSetup**, **PowerRaise**, **PowerLower**, **EndOfProgram**, or the two -token sequence: **User** *user_defined*. These types identify the purpose of each DCSequence block and the environment under which the block is meant to be applied.

13.7 Inherit Constructs

DC parameters are often the same value over many Signals (due perhaps to a common technology across an integrated circuit die, or a common system environment that the device is integrated into, or sometimes due to constraints in the test environment that may not be able to differentiate the Signals), with occasional differences only on specific Signals (due to signal-specific IO constructs). Because of this context, large sections of the DCLevels is often common across Signals except for the little differences.

To address this primary data volume issue, the DCLevels block supports a single level of Inherit information, which is at the top level of the DCLevels statements. At this level, all the constructs of another DCLevels block may be inherited into this block, following the same behavior previously identified with Timing inheritance (in Chapter 10, Advanced Timing). The single DCLevels inherit statement was shown in context with example (13.6), and is repeated here:

```
(13.31)   InheritDCLevels dclevels_block_name ;
```

When a DCLevels block is inherited, all the defined DCLevels statements for each Signal is passed into this block. These statements may be individually replaced by statements found in this block for these Signals.

The next section covers some perspectives on the overall STIL, now that we have introduced several more sections of STIL data.

13.8 STIL Blocks and Data

We have added some additional STIL blocks since the last time we considered the required sequencing of STIL blocks, so let's refresh Figure 6.4 with an update.

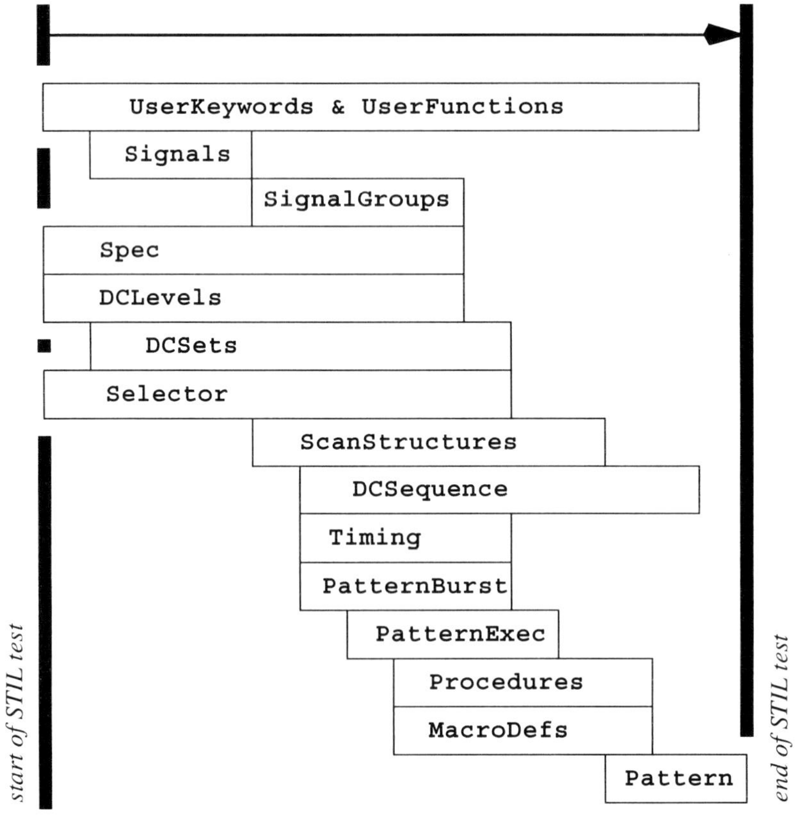

FIGURE 13.1 STIL Block Order

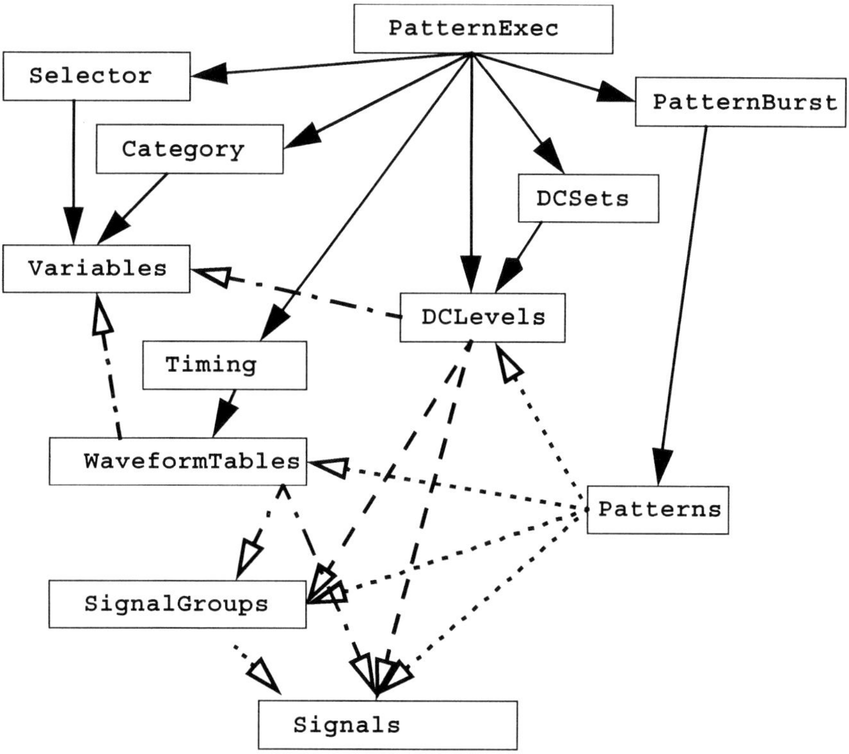

FIGURE 13.2 STIL Data Referencing

Figure 13.2 is fairly confusing, so here's some things to observe:

- There are three sets of lowest level items referenced by other constructs: Signals/SignalGroups, Variables, and WaveformTables.
- There are four sets of data elements that reference these lowest level items: Timing/WaveformTables, Categories&Selectors/Variables, DCLevels/DCSets, and PatternBurst/Patterns.
- The dotted arrows leaving the referencing elements and entering the lowest level elements represent internal references to information contained in the lowest level items. For some of these constructs (notably, the arrows entering the SignalGroups), each arrow also represents a separate resolution process to the information of this item, separate from the other arrows.

CHAPTER 14 More Pattern Constructs

Standard Test Interface Language

This Chapter presents the STIL support for *uncyclized* data. This data is most commonly found in event-based simulation contexts, where the stimulus and response sequences are presented as a (very long) train of events on the Signals. This data is typically not presentable directly to the Test environment without additional processing.

This chapter covers a specific language construct that is *expected* to have very specific and limited application. While it is part of the STIL standard, it is not anticipated to be applied in the test environment, and is not part of the transportation issue central to STIL.

14.1 Event-based Pattern Data

STIL supports direct representation of a stream of events, both stimulus and response, in the Pattern data. This format of data was provided in the Standard to address requirements of ASIC tool kits, which may have a test generation flow from event-based simulation contexts. Event-based simulators operate without an environment of Vectors, cycle boundaries, or repeating event streams. Subsequent processing of this uncyclized data must be performed to create a test that is usable in the test environment.

As such, this capability is not part of the transportation issue that STIL was meant to address (we just identified that this information is typically *not* transported to test environments). But it was part of the standard because without it, some ASIC environments would not be able to have STIL replace their current test generation environments. And that issue was critical for acceptance of the standard.

There are two language constructs to support this context. The first is a statement to establish the baseline context for the other statements. The second statement is another type of assignment statement in a Vector statement. The statements have the form:

```
(14.1)Pattern "x" {
(14.2)    TimeUnits 'time_expression' ;
(14.3)    V {
(14.4)        @ integer_time signal_ref = state ;
(14.5)        @ integer_time { signal_ref = state ; }
(14.6)    }
(14.7)}
```

The **TimeUnits** statement identifies the scale factor associated with all subsequent '@-time' values in the Vectors that may follow.

The '@' statement identifies a single time value offset from the start of the Vector block this statement is found in. The time value is integer only and is multiplied by the TimeUnits expression to resolve to a real time value.

There are two forms of the '@' statement. The first supports a reference to a single named set of signals (an individual signal or one signalgroup, or a signalgroup-on-the-fly), the second statement supports referencing multiple sets of signals under the block construct.

This '@' statement contains a reference to at least all signals that changed value at this time (it may contain references to signals that indicate the same value as last time, but that is redundant information if present). The signals are specified as going to a STIL state_value in this statement.

In this construct there need not be any explicit WaveformTables, no waveforms established, and therefore no referencing of signal behavior through WaveformCharacters. Each event on each signal represents a single transition to another state and not a sequence of events for that signal.

Typically this data is presented in a single Vector in the Pattern block when there is no Timing established. If Timing is present and WaveformTables are referenced in the Pattern block, then the period of each Vector is established by the Period of the current WaveformTable, and there may be multiple Vectors present with uncyclized data. But this situation can occur only if there is Timing present with this information.

STIL Conventions

A STIL test requires mapping of all Signals through WaveformCharacters using established Timing, to be a portable test environment. Uncyclized data does not meet these criteria. Therefore uncyclized data cannot be used for a test environment. However, if all test signals are present in a STIL test with cyclized data present (Timing present and referencing through WaveformCharacters), then the presence of uncyclized data will not affect the test operation. It may be useful to carry this data in a test program to validate additional behaviors or assist in debugging. But when this data is present in parallel with cyclized data *the uncyclized data can and should be ignored in general.*

CHAPTER 15
User Extensions

STIL
Standard
Test Interface
Language

In addition to the Ann statement, STIL includes a robust environment for inserting additional information as part of the test program. This Chapter identifies how to add your own information into a STIL test program, when you should do this, and when you should not.

This chapter covers a feature of the STIL environment that is part of the STIL constructs, and is necessary to be supported by an application that claims support of the language. But these constructs do not (or should not) affect the semantic intent of a STIL test. These construct provide extra features without the foundation of what to do with them — unless your application does understand the additional information.

15.1 UserKeywords statement

STIL supports a level of syntax extension, by allowing you to defined additional keywords to create your own STIL statements. The STIL construct that allows you to define keywords is UserKeywords, and the syntax of this statement is:

```
(15.1)UserKeywords list_of_your_keywords ;
```

The UserKeywords statement can appear at any point amongst the sequence of top-level STIL statements. Typically this statement appears early in the sequence of statements, to define the UserKeywords that may appear up-front in the process and minimize issues.

The STIL UserKeywords statement may occur multiple times, each one defining a single user keyword, or a single UserKeywords statement may be present defining several keywords at once.

Once a UserKeywords statement has declared a user keyword, a STIL statement containing that keyword can appear any place a STIL statement is allowed. This includes inside other STIL blocks, or separately as it's own construct. Once defined, STIL keywords are in effect until the end of the test. There is no scoping of these definitions. They are persistent, they may cross STIL file boundaries and block constructs. They may also be redefined without error. The following code is an example of a UserKeywords context:

```
(15.2)UserKeywords check_its_high ;
(15.3)Timing { WaveformTable "one" { Waveforms {
(15.4)    "in_sigs" { U { '2ns' U; check_its_high; }}
(15.5)}}}
```

There are very few statement types defined for the waveforms block in STIL (although, beware: there are more statements here than you might first remember. In addition to the event statements, Ann statements and Include statements can also appear here). Once defined, UserKeywords can appear in this context as well as within any other STIL context.

One constraint with constructs using STIL keywords is that the contents of the STIL keyword statement must follow STIL syntax rules. The information following a STIL keyword can use either the single-statement form (as shown in the example above), or it may be a STIL block statement containing additional STIL-formatted statements. But all the contents of a STIL UserKeywords statement must follow STIL syntax.

Another constraint of STIL keywords is that they carry *no semantic intent*. No matter what you may think a STIL keyword infers, that inference cannot affect the interpretation of a STIL test. One of the reasons that redefinition of

UserKeywords can occur is because there is no meaning or functionality behind these definitions. Re-declaring the user keyword in multiple UserKeywords statements does not affect the STIL test in any way. This is in contrast to, for instance, macros in some programming languages, where the definition of the construct does define semantic behavior, and redefinition may be an indication of incorrect programming intent.

When a UserKeywords statement uses the STIL block statement, only the outermost statement needs to be declared as a user keyword. Because STIL can not interpret the meaning, nor the contents, of a user keyword block, anything inside a user keyword block has no defined semantic interpretation as part of the STIL standard. Still it must conform to STIL syntax constructs. The following example may help clarify:

```
(15.6)UserKeywords test_attribs ;
(15.7)Signals { "A_control" In {
(15.8)    test_attribs { dont_z; hold_at 1; }
(15.9)}}
```

In this example, only the user keyword test_attribs needed to be declared. The contents of the block defined with test_attribs cannot be used by a general STIL application and therefore the two statements contained in this block do not need to be declared. If these statements occur outside the context of the test_attribs block they would need to be declared as UserKeywords as well.

Ann statements carry the same effective semantic intent as UserKeywords (that is, none), and Ann statements are not constrained to STIL statement constructs when they are used. Both Ann statements and user keyword blocks need to be maintained by STIL parsers and regenerated if the contents of a STIL test are output from an application. There is little motivation to use one construct over the other, unless your application context is extendable and can more easily accommodate the extra information contained with the context of the user keyword construct (in which case, the UserKeyword may be beneficial). If your information is not easily formatted to STIL syntax then the Ann construct may more easily maintain that information.

15.2 Ambiguous Scenarios

The ability to define additional keywords, and create parseable STIL tests with statements using these keywords, can lead to potentially ambiguous situations. UserKeywords in the current standard definition are defined at the top level of STIL blocks, and therefore may appear in any STIL context after they have been declared. Parser error recovery around statements with user keyword statements or blocks (if the statement does not adhere to STIL statement constructs) is more complex as you cannot determine where in the nesting of STIL statements you "should" be with a user keyword.

The following conventions attempt to minimize the potential for misinterpretation.

STIL Conventions

Avoid double-quoting UserKeywords. Quoting UserKeywords makes them appear syntactically as strings, and therefore may be misinterpreted or is at least ambiguous in a context that allows both strings and UserKeywords. For instance:

```
(15.10)UserKeywords "A";
(15.11)Signals {"A" In; // UserKeyword, not a Signal.
```

Consider using a unique naming convention to keep UserKeywords separate from STIL keywords. This can minimize issues with upgrading to future revisions of STIL or supporting additional STIL extensions by eliminating the potential for conflict with STIL keywords in the future. STIL keywords currently maintain the following constraints: they start with an uppercase letter, and they do not contain the underscore character. Both of the examples of UserKeyword application in the section above present UserKeywords that are defined to avoid these conflicts.

15.3 UserFunctions

The last remaining STIL extension statement is the UserFunctions statement. This statement follows the behavior of the UserKeywords statement, and has the same basic syntax:

```
(15.12)UserFunctions list_of_your_functions ;
```

UserFunctions, once defined, may appear in any STIL expression involving real quantities (that is, not signal reference expressions). They provide an extension mechanism for the construction of expressions, but once again, these constructs are provided without semantic definition. That means that the interpretation of these functions is application-specific, and only applications that understand the intent of a defined user function will be able to properly evaluate an expression that contains these functions. Nonetheless, the value in defining UserFunctions is the ability to create a STIL-compliant test for an application that does require these additional functions to be present to properly define expressions.

The intention of the UserFunctions statement is to provide additional function call constructs in the context of expressions. As such, it is anticipated that all uses of that function will operate as if the construct were a function, including parenthesis at the end of the function which may be empty (contain no additional parameters) if the function is not passed arguments in the application. For example:

```
(15.13)UserFunctions test_time stop_signal ;
(15.14)Spec "all" {
(15.15)   Category "threshold" {
(15.16)       "min_tri_time" = '"first_ev" + test_time()';
(15.17)       "max_tri_time" = '"stop_signal("first_ev")';
(15.18)}}
```

Again only the context of the application can interpret these values.

CHAPTER 16

Additional Test Considerations

Standard
Test Interface
Language

We've presented how STIL supports the specification of a wide variety of digital test data. Transporting that data, and using it, in a Test environment is not always straightforward. This Chapter introduces some of the problems here.

Do not look here for answers! At this point in time, the STIL effort is continuing to evolve, and one of the fronts of current development is with Test Resource Constraints. This chapter introduces the fundamental principals for Test Resource Constraint operation, but the details are still in draft form under this Working Group.

16.1 The Concept of Test Constraints

Automated Test Equipment cannot do everything. There are hardware restrictions on many different fronts: the amount of memory available for patterns or timing information, timing restrictions with the accuracy and the flexibility of the timing functions, and functional constraints on what the ATE may or may not be able to do.

Each of the ATE environment restrictions becomes a constraint on what may be represented in a STIL test that is intended to be applied under that ATE environment.

Identifying whether a STIL test is appropriate for a particular ATE environment can be considered a two-step process. First, the contents of the STIL test program need to be investigated, to understand what is present in the program. Then, these constructs need to be measured against the ATE environment and determined whether they violate the ATE capabilities.

Investigating the contents of a STIL test is performed whenever that STIL test is processed. However, maintaining the information about the contents of that test, in a meaningful fashion that could be subsequently interpreted by an application that checks for constraints, requires a standard format for this information to travel between the producer of this data and the consumer.

In addition, specifying the ATE environment constraints in an ATE-independent fashion, such that an independent constraint verifier application could validate these restrictions without the time and effort of attempting to process a STIL test for that ATE environment, also requires a standard definition of these constraints as they apply to the STIL environment.

Both of these efforts are part of the on-going P1450.3 definition. For information about that effort and the current status, contact that Working Group.

CHAPTER 17

Name Mapping

Standard
Test Interface
Language

The process of identifying Signals across the Test environment, or even through the Design environment, may rely on several layers of different name references. The issues here are presented in this Chapter.

Do not look here for answers! At this point in time, the STIL effort is continuing to evolve, and one of the fronts of current development is with Design support, which contains extensions to support name mapping. This chapter introduces the fundamental principals for name mapping, but the details are still in draft form under this Working Group.

17.1 Signal Names

In the test environment, nothing ever goes by one name. Figure 17.1 diagrams one view of this issue. The device Signal name, which we defined at the start of this book, is really a pad on an Integrated Circuit die, with some location reference on that die to identify it. But this pad, in the test environment, is contacted by a loadboard or probe card assembly in the test head, which has another reference to identify which contact point on that board. And this board is plugged into the ATE environment, which establishes a connection with this loadboard contact and defines yet another relationship of loadboard-to-tester channel.

And a packaged part has a separate relationship to the ATE environment.

Ultimately from the perspective of the ATE environment, it is really only the tester channel that matters. It is the tester channel that must provide the stimulus and measure the response of this STIL Signal.

But the tester channel in any particular test environment may be different. Different configurations of ATE environments, different configurations of loadboard or probecards, even different configurations of the IC die may change that final location.

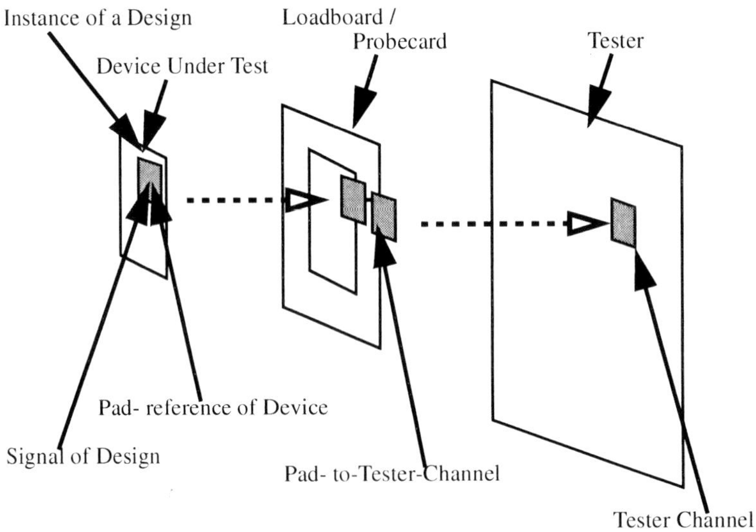

FIGURE 17.1 Names in the Test Environment

Constructs to support this information are being developed as part of the P1450.1 Working Group. Additional information about the names of this Signal, in different contexts, are being defined to support identifying the relationships between these elements as part of the STIL test for this design. These constructs are in draft form only, as are additional extensions to the STIL effort. Contact the Working Group for more information about these features.

Terminology

Standard
Test Interface
Language

Any book covering topics in a specialized field will tend to use the terminology of that field, some of which will be unique to that field and some which may be unique application of terms used elsewhere. Some of the common terms used throughout this book, and the context they are intended for this book, are presented here.

Except for presenting acronyms here, the specialized terms of the STIL language are not listed here. They are covered in the STIL specification itself and in the elaborations found in this book.

ATE refers to Automatic Test Equipment, in particular the consumers of the STIL information defined in this book.

Blank is used to indicate a space (the character used to separate words on this page and generated using the space-bar on a keyboard). Blank is used here to minimize confusion with other meanings of space, and because space takes up too much room.

CAD stands for Computer-Aided Design, and for this book refers specifically to software tools that provide information for test programs.

Contention occurs on bidirectional signals at test when the driver of the device under test, and the driver of the ATE hardware, are both driving active logic values (zero or one). Same-state contention occurs if both drivers are asserting the same logic value (for instance, both are driving a logic zero). Some test environments support or allow some amount of same-state contention. Out-of-state contention occurs when the drivers are in different *active* states, which establishes a short from power to ground through the device, and is never good.

Device refers to an instance of the object being tested. The STIL specification is scoped to support "Integrated Circuits and their Assemblies", therefore when this book refers to a device it is within that context.

DUT refers to Device Under Test. This book uses the term DUT to indicate specifically when the device is being configured under a test context, and uses the term "device" to refer to the device in general, outside the test context.

Lexical refers to the elements and characteristics of a language that define the lowest level of individual entities in the language, or the *tokens*. Functions that operate at the lexical level work between statements in the language and are typically handled by environments that preprocess the language.

Namespace defines the scope of a particular STIL name; the region of STIL constructs over which this name must be unique. Namespaces are particular to specific STIL names or sets of names. In general, any set of names that are used in combination with each other will define a namespace for the entire set.

NRZ refers to Non-Return-to-Zero. This is a waveform term, defined in Table 4.3, "Representative Waveform Styles," on page 67.

Period is used in two distinct contexts in this book:

1. Period refers to the time value associated with a cyclic waveform, and the common cycle timing across all Signals in a WaveformTable.

2. Period refers to the ascii character '.', also called "dot".

Pin is the contact point on the Device Under Test or on the ATE that is the physical connection for the STIL Signal.

PMU refers to the Parametric Measurement Unit on an ATE. This is the specialized hardware used to perform precise measurements of voltage, current, and other parametric characteristics of pins under test.

Scope is used in two contexts:

1. Scope refers to the defined area of support for a Standard. The scope of a Standard identifies what the Standard is defined to support; by implication if something is not defined in the scope then it is outside the Standard.
2. Scope is also applied to the region of a STIL test over which a particular name in a particular construct is visible. Within this scope, each name must be unique because only one individual item can be referenced from this name. In STIL, the concept of a namespace is used to represent this scope in combination with the specific constructs (see namespace).

Signal refers to the STIL construct that references the port of a Device Under Test, through which test information travels.

STIL test is a general reference to a sequence of STIL data constructs such that it defines a complete set of STIL information necessary to run that test in an application - be that application a test environment, a simulation context, or a STIL data processing tool. A STIL **test**, when running, always starts with a PatternExec block, and generally ends with a Pattern block, although the sequence of STIL data constructs starts with a STIL version statement followed shortly by the Signals block.

String is a general reference to a sequence of ascii characters. It can be considered to be a generalization of the concept of a word - sometimes, it may be a word, but at other times it may just be a sequence of characters. When looking at a sequence of WaveformCharacters, the perspective of a "string" of characters is generally pretty apparent.

Syntactic refers to the statements of a language; the sequence of tokens that defines a meaningful construct in that language.

Tester Channel is the hardware behind the tester pin, that supports the test information and implements the functionality to support the test operation.

Time-zero refers to the start of each STIL Vector in a test program. This term may have different definitions in different ATE environments, in this book it refers to the start of a Vector.

Token is a term used in the context of parsing languages. It identifies a discrete textual unit, or a word, of the language. in this book a token refers to a contiguous set of characters which may or may not be a complete word depending on the concatenation operator as defined in Section 3.1.2, page 36.

WFC stands for WaveformCharacter, which is a single-character reference associated with a waveform definition and is the mechanism to reference these waveforms in the Pattern block. WaveformCharacters are defined in Chapter 4, Timing, and are used in Chapter 5, Patterns.

WFT is short for WaveformTable. This is not generally used in the text but many examples may reference this term as a shorthand notation.

Whitespace is the construct that separates words. For STIL, explicit whitespace is blanks, tabs, and carriage returns. In addition, implicit whitespace (which defines the end of one word and therefore the start of the next word even without one of these explicit characters) is identified in Table 3.1, "Characters that Terminate STIL Tokens," on page 36.

Bibliography

Standard
Test Interface
Language

[1] IEEE Standards Association, *IEEE Standard Test Interface Language (STIL) for Digital Test Vectors*, IEEE Standard 1450-1999.

[2] IEEE Standards Association, *IEEE Standard for Digital Test Interchange Format (DTIF)*, IEEE Standard 1445-1998.

[3] IMS Corporation, *Waveform Generation Language*.

[4] IEEE Standards Association, *IEEE Standard for Verilog® Hardware Description Language*, IEEE Standard 1364-2001.

[5] IEEE Standards Association, *IEEE Standard for Extensions to Standard Test Interface Language (STIL) (IEEE Std 1450™-1999) for DC Level Specification*, IEEE Standard 1450.2™-2002.

STIL Syntax Summary

Standard
Test Interface
Language

The following pages outline the complete set of STIL syntax constructs, both for current standards and proposed constructs from working drafts. *Please be aware that the proposed constructs are subject to change.*

Also be aware that these syntax summaries reflect language constructs only. The presentation in this book has been at the functional level. There is not a direct correlation of every syntax element presented in this section, with a reference in this book. There is a closer correlation of this representation with the IEEE standards publication, and it is recommended that the official standards be consulted to resolve questions.

STIL Reference Guide

STIL.0 - Basic STIL; approved by IEEE in 1999
STIL.1 - Design extension; working draft D16, June 19, 2003
STIL.2 - DC Levels; approved by IEEE in 2002
STIL.3 - Tester Resource Constraints extension; working draft D8, Apr 11, 2003
STIL.4 - Flow extension; *not included*
STIL.5 - Test Method extension; *not included*
STIL.6 - CTL extension (Core Test Language)

Lexical Environment

Simple statement construct:
Keyword (OPTIONAL_TOKENS)***;**

Block statement construct:
Keyword (OPTIONAL_TOKENS)* **{** (OPTIONAL_STATEMENTS)* **}**

Whitespace:
space tab newline_character statement_terminators

Comments:
// LINE COMMENT - *line comments are terminated by newline*
/* BLOCK
COMMENT */ - *block comments may span multiple lines*

IEEE Std. 1450-1999

```
STIL 1.0;
Header {
   ( Title "TITLE_STRING"; )
   ( Date "DATE_STRING"; )
   ( Source "SOURCE_STRING"; )
   ( History { } )
}

Include "file_name" (IfNeed <blocktype>);
Ann {* ANNOTATION *}

UserKeywords (USER_KEYWORDS)+;
UserFunctions (USER_KEYWORDS)+;


Signals {
   ( SIG_NAME < In | Out | InOut | Supply | Pseudo >; )*
   ( SIG_NAME < In | Out | InOut | Supply | Pseudo > {
      ( Termination < TerminateHigh | TerminateLow |
TerminateOff | TerminateUnknown > ; )
      ( DefaultState < U|D|Z|ForceUp| ForceDown | ForceOff > ; )
      ( Base < Hex | Dec > WAVEFORM_CHARACTER_LIST ; )
      ( Alignment <MSB | LSB> ; )
      ( ScanIn (DECIMAL_INTEGER) ; )
      ( ScanOut (DECIMAL_INTEGER) ; )
      ( DataBitCount DECIMAL_INTEGER ; )
      } )*
   }
```

```
SignalGroups (DOMAIN_NAME) {
   ( GROUPNAME = sigref_expr; )*
   ( GROUPNAME = sigref_expr { ( Termination
      < TerminateHigh | TerminateLow | TerminateOff |
TerminateUnknown > ; )
      ( DefaultState < U | D | Z | ForceUp | ForceDown | ForceOff
> ; )
      ( Base < Hex | Dec > WAVEFORM_CHARACTER_LIST ; )
      ( Alignment <MSB | LSB> ; )
      ( ScanIn (DECIMAL_INTEGER); )
      ( ScanOut (DECIMAL_INTEGER); )
      ( DataBitCount DECIMAL_INTEGER; )
      } )*
}            // end of SignalGroups

PatternExec (PAT_EXEC_NAME) {
   ( Category CATEGORY_NAME ; )*
   ( Selector SELECTOR_NAME ; )*
   ( Timing TIMING_NAME ; )
   ( PatternBurst PAT_BURST_NAME ; )
}            // end of PatternExec

PatternBurst PAT_BURST_NAME {
   ( SignalGroups GROUPS_DOMAIN ; )*
   ( MacroDefs MACROS_DOMAIN ; )*
   ( Procedures PROCEDURES_DOMAIN ; )*
   ( ScanStructures SCAN_NAME ; )*
   ( Start PAT_LABEL ; ) ( Stop PAT_LABEL ; )
   ( Termination { ( sigref_expr < TerminateHigh |
TerminateLow | TerminateOff | TerminateUnknown > ; )*
   } )  // end Termination
   ( PatList {
      ( PAT_NAME_OR_BURST_NAME ; )*
      ( PAT_NAME_OR_BURST_NAME {
         ( SignalGroups GROUPS_DOMAIN ; )*
         ( MacroDefs MACROS_DOMAIN ; )*
         ( Procedures PROCEDURES_DOMAIN ; )*
         ( ScanStructures SCAN_NAME ; )*
         ( Start PAT_LABEL ; )
         ( Stop PAT_LABEL ; )
         ( Termination { ( sigref_expr <TerminateHigh |
TerminateLow | TerminateOff |TerminateUnknown>;)*
         } )   // end Termination
      } )*     // end of Pat_name_or_Burst_name
   } )+        // end of PatList
}              // end of PatternBurst

Timing ( TIM_DOMAIN_NAME ) {
   ( SignalGroups GROUPS_DOMAIN ; )*
   ( WaveformTable WFT {
      ( Period time_expr; )
      ( InheritWaveformTable (TIM.)WFT ; )*
      Waveforms {
         ( sigref_expr (WAV) {
            ( InheritWaveform ((TIM.)WFT.)WAV ; )*
            ( WFC {
               ( InheritWaveform (((TIM.)WFT.)WAV.)WFC ; )*
               (( EVENT_LABEL: ) ( time_expr ) ( event ) ; )*
               } )* // end wfc
```

```
        ( WFC_LIST {
          ( InheritWaveform (((TIM.)WFT.)WAV.)WFC; )*
          (( EVENT_LABEL: ) ( time_expr )
                ( <event_list ( [EVENT_NUM] ) | event > ) ; )*
          } )* // end wfc_list
      ( <WFC | WFC_LIST> { (( EVENT_LABEL: ) ( time_expr ) (\rN)
          < SUB_WF_LAB ; | SUB_WF_LAB[N] ;
          | SUB_WF_LAB[#] ; > )* } )*
        } )*  // end wfc | wfc_list
      }       // end sigref_expr
    } )*      // end Waveforms
  }           // end Waveforms
    ( SubWaveforms {
      ( SUB_WF_LAB: Duration time_expr {
        (( EVENT_LABEL: )( time_expr )
              ( <EVENT_LIST ( [EVENT_NUM] ) | event > ) ; )*
      } )*    // end of Duration
    } )       // end of SubWaveforms
  })          // end of WaveformTable
}             // end of Timing
```

```
// This form of the Spec defines Variables in a given Category
Spec (SPEC_NAME) {
  ( Category CAT_NAME {
    ( VAR_NAME = time_expr; )*// define only the Typ value
    ( VAR_NAME {
      (Min time_expr;) (Typ time_expr;) (Max time_expr;) } )*
  } )+      // end of Category
}           // end of Spec
```

```
// This form of the Spec defines Category values for a given Variable
Spec (SPEC_NAME) {
  ( Variable VAR_NAME {
    ( CAT_NAME = time_expr; )*// defines only the Typ value
    ( CAT_NAME {
      (Min time_expr;) (Typ time_expr;) (Max time_expr;) } )*
  } )+      // end of Variable
}           // end of Spec
```

```
Selector SELECTOR_NAME {
  ( VAR_NAME < Min | Typ | Max | Meas >; )* }
```

```
ScanStructures (SCAN_NAME) {
  ( ScanChain CHAIN_NAME {
    ScanLength DECIMAL_INTEGER;
    ( ScanOutLength DECIMAL_INTEGER; )
    ( ScanCells CELLNAME-LIST ; )
    ( ScanIn SIGNALNAME; )
    ( ScanOut SIGNALNAME; )
    ( ScanMasterClock SIGNALNAME-LIST ; )
    ( ScanSlaveClock SIGNALNAME-LIST; )
    ( ScanInversion < 0 | 1 >; )
  } )*      // end of ScanChain
}           // end of ScanStructures
```

```
Pattern PATTERN_NAME {
   ( LABEL : )
   ( TimeUnit 'TIME_DEF'; )
   ( V(ector) { ( cyclized-data )* ( non-cyclized-data )* } )
   ( W(aveformTable) NAME ; )
   ( C(ondition) { ( cyclized-data )* ( non-cyclized-data )* } )
   ( Call PROCNAME ; )
   ( Call PROCNAME { ( scan-data )* ( cyclized-data )* ( non-
cyclized-data )* } )
   ( Macro MACRONAME ; )
   ( Macro MACRONAME { ( scan-data )* ( cyclized-data )* ( non-
cyclized-data )* } )
   ( Loop LOOPCNT { ( pattern-statements )* } )
   ( MatchLoop < LOOPCNT | Infinite > {
      (pattern-statements)+
      ( BreakPoint { (pattern-statements)+ } )
        } ) // end MatchLoop
   ( Goto LABELNAME ; )
   ( BreakPoint; )
   ( BreakPoint { ( PATTERN-STATEMENTS )* } )
   ( IddqTestPoint; )
   ( Stop ; )
   ( ScanChain CHAINNAME ; )
}          // end of Pattern

Procedures ( PROCEDURE_DOMAIN_NAME ) {
   ( PROCEDURE_NAME {
      ( PATTERN-STATEMENT )*
      ( Shift { (pattern-statements)+ } )
   } )*
}
```

```
MacroDefs ( MACRO_DOMAIN_NAME ) {
   (  MACRO_NAME {
      ( PATTERN-STATEMENT )*
      ( Shift { (pattern-statements)+ } )
   } )*
}
```

vec_data:

\w — WaveformChar format (\w0100 xxxx qqqQQQ).

\rN XXX — Repeat the next group of characters N times (\r60 01).

\h — Hexadecimal format (\h E4B2).

\hCHARS — Hexadecimal format with WFC (\h01 69BC).

\d — Decimal format (\d 79)

\dCHARS — Decimal format with WFC association (\dHL 375).

\lN — Shift data passed in procedure or macro calls, or multiple-bit data passed into waveforms that contain square-bracket constructs.

— replaces the parameter with the next argument value from the call.

% — replaces the parameter with the current argument value from the call.

IEEE P1450.1, Draft 16, June 19, 2003

```
STIL 1.0 { Design D16 ; }

Variables ( VARS_BLOCK_NAME ) {
  ( IntegerConstant CONST_NAME = DECIMAL_INTEGER ; )*
  ( Integer VAR_NAME ; )*
  ( Integer VAR_NAME {
    ( Usage Test ; )
    ( InitialValue logical_expr ; )
  } )*        // end of Integer
  ( SignalVariable VAR_NAME ; )*
  ( SignalVariable VAR_NAME   {
    ( Base < Hex | Dec> WFC_LIST ; )
    ( Alignment < MSB | LSB > ; )
    ( InitialValue vec_data ; )
  } )*        // end SignalVariable
  ( WFCConstant CONST_NAME = WFC_LIST ; )*


Signals {
  ( SIG_NAME < In | Out | InOut | Supply | Pseudo > {
    ( WFCMap {
      ( FROM_WFC -> TO_WFC( TO_WFC )+ ; )*
      ( FROM_WFC1 FROM_WFC2 -> TO_WFC ; )*
    } )       // end of WFCMap
  } )*
}             // end of Signals

SignalGroups {
  ( GROUPNAME = sigref_expr {
    ( WFCMap {
      ( FROM_WFC -> TO_WFC( TO_WFC )+ ; )*
      ( FROM_WFC1 FROM_WFC2 -> TO_WFC ; )*
    } )         // end of WFCMap
  } )*
}               // end of SignalGroups


PatternBurst PAT_BURST_NAME {
  ( Variables VARIABLES_DOMAIN ; )*
  ( Fixed { (cyclized-data)* } )
  ( PatList {
    ( PAT_NAME_OR_BURST_NAME {
      ( Variables VARIABLES_DOMAIN ; )*
      ( Fixed { (cyclized-data)* } )
      ( If boolean_expr ; )
      ( While boolean_expr ; )
    } )*          // end of pat_name_or_burst_name
  } )*            // end of PatList
  ( PatSet {
    ( PAT_NAME_OR_BURST_NAME ; )*
    ( PAT_NAME_OR_BURST_NAME {
      ( Variables VARIABLES_DOMAIN ; )*
      ( Fixed { (cyclized-data)* } )
    } )*          // end of pat_name_or_burst_name
  } )*            // end of PatSet
```

```
  ( ParallelPatList ( SyncStart | Independent | LockStep ) {
     ( PAT_NAME_OR_BURST_NAME ; )*
     ( PAT_NAME_OR_BURST_NAME {
        ( Variables VARIABLES_DOMAIN ; )*
        ( Fixed { (cyclized-data)* } )
        ( Wait ; )
        ( Extend ; )
        ( If boolean_expr ; )
        ( While boolean_expr ; )
     } )*          // end of pat_name_or_burst_name
  } )*             // end of ParallelPatList
}                  // end of PatternBurst

Timing TIMING_NAME { ( Variables VARIABLES_DOMAIN ; )* }

ScanStructures ( SCAN_STRUCT_NAME ) {
  ( ScanChain CHAIN_NAME {
     ( ScanLength integer_expr ; )
     ( ScanOutLength integer_expr ; )
     ( ScanEnable logical_expr ; )
     ( ScanCells {
        ( scan_cell_name ; )*
        ( scan_cell_name {
           ( (If boolean_expr) CellIn STATE-ELEMENT-LIST ; )*
           ( (If boolean_expr) CellOut STATE-ELEMENT ; )*
        } )*       // end scan_cell_name
     } )           // end ScanCells
  } )*             // end ScanChain
  ( ScanChainGroups {
     ( GROUPNAME { (GROUP_OR_CHAIN_NAME ; )* } )*
  } )              // end ScanChainGroups
}                  // end ScanStructures
ScanCellGroups ( DOMAIN_NAME ) {
  ( cellgroupname = cellref_expr ; )*
  ( cellgroupname = cellref_expr {
     ( Base < Hex | Dec > WAVEFORM_CHARACTER_LIST ; )
     ( Alignment <MSB | LSB> ; )
     ( DataBitCount DECIMAL_INTEGER ; )
  } )              // end of cellgroupname
}                  // end of ScanCellGroups

Environment (ENV_NAME) {
  ( InheritEnvironment ENV_NAME ; )*
  ( NameMaps (MAP_NAME) {
     InheritNameMap ENV_NAME(.MAP_NAME) ;
     ( ScanCells { (SCAN_CELL_NAME "MAP_STRING" ; )* } )*
     ( Signals { (SIG_NAME "MAP_STRING" ; )* } )*
     ( SignalGroups (DOMAIN_NAME) { (GROUP_NAME
"MAP_STRING" ; )* } )*
     ( Variable { (VAR_NAME "MAP_STRING" ; )* } )*
     ( AllNames { (ANY_NAME "MAP_STRING" ; )* } )*
  } )*             // end of NameMaps
  ( FileReference "file_path_name" ; )*
  ( FileReference "file_path_name" {
     Type FILE_TYPE ;
     Format FILE_FORMAT ;
     Version "VERSION_NUMBER" ;
  } )*             // end of FileReference
  (ENVIRONMENT_DEFINED_STATEMENTS)*
}                  // end of Environment
```

sigref_expr = *\readback_function* SIGNAL_NAME;
sigref_expr = *\readback_function* SIGNAL_GROUP_NAME;
sigref_expr = *\readback_function* SIGNAL_VARIABLE_NAME;
sigref_expr = *\readback_function* WFCCONSTANT_NAME;

If '*\readback_function* SIGNAL_NAME **==** *vec_data*' { }
If '*\readback_function* SIGNAL_NAME **==** *event*' { }
If '*\readback_function* SIGNAL_GROUP_NAME **==** *vec_data*' { }
If '*\readback_function* SIGNAL_GROUP_NAME **==** *event*' { }
If '*\readback_function* SIGNAL_VARIABLE_NAME **==** *vec_data*' { }

readback_function:
\C SIGNAL-OR-GROUP-NAME — return last compare event/event-list
\D SIGNAL-OR-GROUP-NAME — return the last drive event/event-list
\E SIGNAL-OR-GROUP-NAME — return the last expect event/event-list
\S SIGNAL-OR-GROUP-NAME — return last wfclist using (s or S) event
\U SIGNAL-OR-GROUP-NAME — return last undefined event/event-list
\W SIGNAL-OR-GROUP-OR-SIGNALVARIABLE-OR-WFCCONSTANT-NAME — return the last wfc/wfc-list

Pattern PATTERN_NAME **{**
(**If** *boolean_expr* **{** (PATTERN_STATEMENTS)* **}**)
(**Else** **{** (PATTERN_STATEMENTS)* **}**)
(**While** *boolean_expr* **{** (PATTERN_STATEMENTS)* **}**)
(**F(ixed)** **{** (*cyclized-data*)* (*non-cyclized-data*)* **}**)
(**E(quivalent)** *sigref_expr* **;**)
(**X** tag **;**)
(**X** (tag) **{**
(**Cycle** *integer* **;**)
(**Iteration** *integer* **;**)
(**PatternOffset** *integer* **;**)
(**TagOffset** *integer* **;**)
}) // *end X*
(**Loop Data** **{** (PATTERN-STATEMENTS)* **}**)
(**Loop** < *boolean_expr* | *integer_expr* > **{** (PATTERN_STATEMENTS)* **}**)
(**ActiveScanChains** (GROUP_OR_CHAIN_NAME)* **;**)
(**AllowInterleave** *sigref_expr* **;**)
(**BreakPoint** **{}**)
} // *end Pattern*

Pragma NAME **{*** ... APPLICATION DEPENDANT SYNTAX ... ***}**

IEEE Std. 1450.2-2002

```
STIL 1.0 { DCLevels 2002 ; }

PatternExec ( PAT_EXEC_NAME ) {
   ( DCLevels ( DC_LEVELS_NAME ) ; )
   ( DCSets ( DC_SETS_NAMES ) ; )
}

( DCLevels ( DC_LEVELS_NAME ) {
   ( InheritDCLevels DC_LEVELS_REF ; )*
   ( SignalGroups GROUPS_DOMAIN ; )*
   ( sigref_expr { // Signals of type In, Out, InOut, Supply
   (VIH (dc_expr)+;)                (VIL (dc_expr)+;)
   (VICM dc_expr;)                  (VID dc_expr;)
   (VIHD dc_expr;)                  (VILD dc_expr;)
   (ForceHi;)                       (ForceLo;)
   (InitHi;)                        (InitLo;)
   (VIHSlew dc_expr;)               (VILSlew dc_expr;)
   (VOH (dc_expr)+;)                (VOL (dc_expr)+;)
   (VOCM dc_expr;)                  (VOD dc_expr;)
   (VOHD dc_expr;)                  (VOLD dc_expr;)
   (IOH dc_expr;)                   (IOL dc_expr;)
   (LoadVRef dc_expr;)              (ClampHi dc_expr;)
   (ClampLo dc_expr;)               (TermVRef dc_expr;)
   (VForce dc_expr;)                (IClamp dc_expr;)
   (IForce dc_expr;)                (VClamp dc_expr;)
   (ResistiveTermination dc_expr;)
   } )*              // end of sigref_expr
} )*                 // end of DCLevels

( DCSets ( DC_SET_NAME ) {
   ( DCLevels ( DC_LEVELS_REF ) ; )*
} )*

( DCSequence ( < InitialSetup | PowerRaise | PowerLower |
EndOfProgram | User USER_DEFINED >) {
   ( SignalGroups GROUPS_DOMAIN ; )*
   ( time_expr sigref_expr {
      ( Connect ( < Supply | PMU | Driver | Comparator| Load |
Termination | Clamp > )* ; )*  // Physical connection (e.g., relay)
      ( Disconnect ( < Supply | PMU | Driver | Comparator | Load
| Termination | Clamp > )* ; )*
      ( Apply ( dc_expr ) ; )* // optional voltage or current value
      ( Ramp time_expr (dc_expr) ; )* // ramp duration
   } )*              // end time_expr
} )*                 // end DCSequence
```

IEEE P1450.3, Draft 8, Apr 11 2003

```
STIL 1.0 { TRC D08; }

Resource (TESTER_IDENTIFIER)+ ; // list of target testers
   < (RESOURCE_ID)+ >
}

Environment (ENV_NAME) {
   ( NameMaps (MAP_NAME) {
      ( Signals { (SIG_NAME "MAP_STRING"; )* } )*
   }              // end NameMaps
}                 // end Environment

Environment (ENV_NAME) {
   ( TRC (TRC_NAME) {
      ( Category (CAT_NAME)+ ;)
      ( Selector (SEL_NAME)+ ;)
      ( NameChecks (NAM_CHK_NAME) { } )*
      ( PatternCharacteristics PAT_CHAR_NAME { } )*
      ( PeriodCharacteristics PER_CHAR_NAME { } )*
      ( SignalCharacteristics (signal_attributes)+ { } )*
      ( SystemCharacteristics {
         ( MultipleDevices integer_expr;)
         ( MultipleSites integer_expr;)
      } )
      ( Usage < Constraints | PatternReport > ; )
      ( WaveformCharacteristics WAV_CHAR_NAME { } )*
      ( WaveformDescriptions WAV_DESC_NAME (Explicit) { } )*
   })* // end TestResourceConstraints
} // end Environment
```

```
SignalCharacteristics
   (<Data| Clock | Scan| In | Out| InOut | SplitIO| Asynchronous |
User USER_DEFINED>)+ {
      ( DCResourceCharacteristics DC_CHAR_NAME; )
      ( FanOut integer_expr; )
      ( InOut < WithinCycle | OnCycleBoundary | Static >;)
      ( MaxScanMemory integer_expr;)
      ( MaxCaptureMemory integer_expr (<FailOnly|Result>)+ ;)
      ( MaxScanChainLength integer_expr ; )
      ( MaxSignals integer_expr;)
      ( MaxVectorMemory integer_expr;)
      ( PatternCharacteristics PAT_CHAR_NAME ;)
      ( PeriodCharacteristics PER_CHAR_NAME (<Synchronous |
Asynchronous>)+ ;)
      ( UndrivenInOut < Yes | No >;)
      ( WaveformCharacteristics WAV_CHAR_NAME ;)
      ( WaveformDescriptions WAV_DESC_NAME ;)
   } // end SignalCharacteristics

DCResourceCharacteristics (DC_RESOURCE_NAME) {
   ( PerPinConfiguration
      (<Supply | PMU | Driver | Comparator | Load | Termination
| Clamp>)+; )
   ( DifferentialConfiguration (<Driver | Comparator>)+;)
   ( NumberTesterChannels integer_expr;)
   ( DCLimits < VIH | VIL | VICM | VID | VIHD | VILD |
VIHSlew | VILSlew| VOH | VOL | VOCM | VOD | VOHD |
VOLD | IOH | IOL | LoadVRef | ClampHi | ClampLo |
ResistiveTermination | TermVRef | VForce | IClamp | IForce |
VClamp >
      { ( Min dc_expr; ) ( Max dc_expr;) } )*
```

```
   ( NumberLevels (<VIH | VIL | VOH | VOL>)+ integer_expr ; )*
   ( NumberDCLevels integer_expr ; )
   ( NumberDCLevels integer_expr { SignalsPer integer_expr ; }
}

PeriodCharacteristics (PER_CHAR_NAME) {
   (Accuracy time_expr;)
   (MaxPeriods integer_expr;)
   (MaxPeriodGenerators integer_expr (< Dynamic | Static >) ; )
   (MaxPeriodGenerators integer_expr (< Dynamic | Static >) {
      ( SignalsPer integer_exp; )
   } // end MaxPeriodGenerators
   ( PeriodSelectMemory integer_expr; )
   ( PeriodSelectMemory integer_expr {
      ( SignalsPer integer_exp; )
   } ) // end PeriodSelectMemory
   ( TimeLimits boolean_expr; )
   (Resolution time_expr;)
} // end PeriodCharacteristics

WaveformCharacteristics (WAV_CHAR_NAME) {
   ( Accuracy <Edge | EdgeToEdge> time_expr;)*
   ( CompareEvents
      (<H|L|X|V|T|h|l|x|v|t>(/<H|L|X|V|T|h|l|x|v|t>)*)+
      (integer_expr (integer_expr)) ; )*
   ( DriveEvents
      (<U|D|Z|P>(/<U|D|Z|P>)*)+
      (integer_expr (integer_expr)) ; )*
// use the following syntax to define MaxShapes
   ( FormatSelect < In | Out | InOut > {
      ( MaxShapes integer_expr (< Dynamic | Static >); )
      ( MaxShapes integer_expr (< Dynamic | Static >) {
        ( SignalsPer integer_exp; )
      }              // end MaxShapes
   }                 // end FormatSelect (with MaxShapes)
// use following syntax in place MaxShapes (above)
   ( FormatSelect < In | Out | InOut > {
      ( MaxTimeSets integer_expr (< Dynamic | Static >) ; )
      ( MaxTimeSets integer_expr (< Dynamic | Static >) {
        ( SignalsPer integer_exp; )
        ( PerTimingGenerator; )
      } )            // end MaxTimeSets
      ( MaxTimingGenerators integer_expr (<Dynamic|Static>) ; )
      ( MaxTimingGenerators integer_expr (< Dynamic|Static >) {
        ( SignalsPer integer_exp; )
      } )            // MaxTimingGenerators
      ( MaxData <Drive|Compare|DriveCompare> integer_expr
(<Dynamic|Static>);
      ( MaxData <Drive|Compare|DriveCompare> integer_expr
(<Dynamic|Static> {
        ( SignalsPer integer_exp; )
      } )            // end MaxData
      ( MaxIO integer_expr integer_expr (< Dynamic | Static >); )
      ( MaxIO integer_expr integer_expr (< Dynamic | Static >) {
        ( SignalsPer integer_exp; )
      } )            // end MaxIO
      ( MaxMask integer_expr (< Dynamic | Static >); )
      ( MaxMask integer_expr (< Dynamic | Static >) {
        ( SignalsPer integer_exp; )
      } )            // end MaxMask
   })*               // end FormatSelect
   ( MaxEdgeTime time_expr;)
```

```
  ( MinCompareWindow time_expr;)
  ( MinCompareToDriveOn time_expr;)
  ( MinEdgeReTrigger time_expr (<Drive | Compare>)* ;)*
  ( MinDriveOffTime time_expr;)
  ( MinDriveOffToCompare time_expr;)
  ( MinDriveOnTime time_expr;)
  ( MinDrivePulse time_expr;)
  ( ReplicateSubWaveform integer_expr; )
  ( Resolution time_expr;)
  ( TimeLimits boolean_expr; )
  ( WaveformSelectMemory integer_expr ; )
  ( WaveformSelectMemory integer_expr {
    ( SignalsPer integer_exp; )
    ( SelectWithPeriod; )
  } )        // end WaveformSelectMemory
}            // end WaveformCharacteristics

WaveformDescriptions (WAV_DESC_NAME) (Explicit) {
  ( < In | Out | InOut > WFNAME {
    ( NumberData integer_expr ; )
    ( NumberIO integer_expr ; )
    ( NumberMask integer_expr ; )
    ( NumberPeriods integer_expr;)
    ( NumberShapes integer_expr ; )
    ( NumberSignals integer_expr;)
    ( NumberTimeSets integer_expr ; )
    Shape {
      ( (TRC_LABEL:) ( time_expr ) < event | event_list> ;)*
    } )*       // end Shape
  } )*         // end < In | Out | InOut >
}              // end WaveformDescriptions
```

```
PatternCharacteristics (PAT_CHAR_NAME) {
  ( Base < Hex | Dec | HexDec > integer;)
  ( BreakPoint < Yes | No >;)
  ( ConditionalStatements < Yes | No > ;)
  ( CoreUsageReady < Yes | No > ;)
  ( DeltaChangeVectorData < Yes | No >;)
  ( GoTo < Yes | No >;)
  ( IddqTestPoints < Yes | No >;)
  ( InstructionCharacteristics {
    ( AllowedWhen boolean_expr; )
    ( AppliesTo (< Condition | GoTo | Loop | MatchLoop | Call |
Shift >)+ ;)
    ( LoopCharacteristics LOOP_NAME {
      ( Infinite; )
      ( MaxIteration integer_expr; )
      ( MaxLength integer_expr; )
      ( MaxNest integer_expr; )
      ( MinIteration integer_expr; )
      ( MinLength integer_expr; )
      ( MinTimeAfterMatch time_expr; )
      ( MinVectorsAfterMatch integer_expr; )
      ( VectorModulus integer_expr; )
    } )* // end LoopCharacteristics
    ( MinVectorsAfter integer_expr; )
    ( MinVectorsBefore integer_expr; )
    ( MinVectorsBetween integer_expr; )
  } )* // end InstructionCharacteristics
  ( Macro < Yes | No | WithParameters >;)
  ( MaxRunTime time_expr;)
  ( MaxVectors integer_expr;)
  ( MultiBitData (< InWaveforms | InPatterns | No >)+ ;)
```

```
( NonCyclized < Yes | No >;)
( NumberCaptureCycles (integer_expr ( integer_expr)) ;)
( NumberPatternUnits integer_expr ;)
( NumberVectorsPerShift (integer_expr) ( integer_expr);)
( PatternVariables (< Integer | IntegerConstant |
SignalVariable | Spec | All >* ;)
( ProcedureCalls < Yes | No| WithParameters > ;)
( Shift < Yes | No >;)
( STILPatterns < Yes | No >;)
( VectorModulus integer_expr; )
} // end PatternCharacteristics

PatternInformation {
  < Pattern | PatternBurst > PAT_OR_BURST_NAME {
  ( PatternCharacteristics { } )
}
NameChecks NAM_CHECKS_NAME {
  ( Contents (STIL_BLOCK_NAME)+ ; )
  ( Block STIL_BLOCK_NAME ; )*
  ( CharacterContent 'regular_expr' ; )
  ( Length integer_expr ; )
  ( Scope (STIL_BLOCK_NAME)+ ; )
}
```

IEEE P1450.6, Draft 18, Feb. 12 2002

```
STIL 1.0 { CTL  D18 ; }

(CoreType CORE_TYPE_NAME {
    ( CoreEnvironment CORE_ENV_NAME; )
    ( InheritCoreType CORE_TYPE_NAME {
        CoreInstance (CORE_INSTANCE_NAME)+ ;
    } )* // end InheritCoreType
    ( ScanStructures (SCAN_STRUCT_NAME) {
        core_scan_structure_definitions
    } )* // end ScanStructures
    ( Signals {
        core_signal_definitions
    } ) // end Signals
    ( SignalGroups (DOMAIN_NAME) {
        core_signal_group_definitions
    } )* // end SignalGroups
  })*// end CoreType

  ( CoreInstance CORE_TYPE_NAME {
    ((CORE_INSTANCE_NAME)+;)
    (CORE_INSTANCE_NAME { PatternBurst (BURST_NAME)+;})*
  })*

ScanCellGroups (DOMAIN_NAME) {
  ( CELLGROUPNAME = cellref_expr; )*
  ( CELLGROUPNAME = cellref_expr {
    ( Base < Hex | Dec > WAVEFORM_CHARACTER_LIST ; )
    ( Alignment <MSB | LSB> ; )
    ( DataBitCount DECIMAL_INTEGER; )
```

```
    } ) // end cellgroupname
} // end ScanCellGroups

( FileReference "FILE_PATH_NAME"; )*
( FileReference "FILE_PATH_NAME" {
   Type file_type ;
   Format < pattern_file_format | design_file_format |
layout_file_format
       | fault_list_file_format | script_file_format | doc_file_format > ;
   Version "VERSION_NUMBER" ;
} )*                // end FileReference

file_type =
   < Pattern| Design
   | Layout| FaultList
   | Script| Documentation
   | User USER_DEFINED >

pattern_file_format =
   < STIL| WGL
   | Verilog| VHDL
   | VCD| User USER_DEFINED >

design_file_format =
   < CTL| EDIF
   | Verilog| VHDL
   | User USER_DEFINED >

layout_file_format =
   < GDSII| User USER_DEFINED >

fault_list_file_format =
   < DTIF| User USER_DEFINED >

script_file_format =
   < AWK| Perl
   | TCL| User USER_DEFINED >

doc_file_format =
   < HTML| PDF
   | Postscript| RTF
   | Text| User USER_DEFINED >

test_mode_enum =
   < Bypass| Debug
   | ExternalTest| ForInheritOnly
   | InternalTest| Isolate
   | Normal| PowerOff
   | PowerOn| PreLoad
   | Quiet| Repair
   | Sample| Transparent
   | User USER_DEFINED >
cell_enum = <
          WC_<D|S>#(<D|S>U)(_S) // refer to P1500-CTAG for
semantic definition
          | User USER_DEFINED >

e.g., WC_S1, WC_S1_S, WC_D1, WC_D2DU, WC_D2SU_S
```

```
data_type_enum =
   < Asynchronous| ChipSelect
   | Clear| ClockEnable
   | Constant| Functional
   | In| InOut
   | Indicator| InOutControl
   | Oscillator| Out
   | OutDisable| OutEnable
   | MasterClock| MemoryAddress
   | MemoryData| MemoryRead
   | MemoryWrite| Programming
   | Regular| Reset
   | ScanDataIn| ScanDataOut
   | ScanEnable| ScanMasterClock
   | ScanSlaveClock| Set
   | SlaveClock| Synchronous
   | TestAlgorithm| TestClock
   | TestControl| TestData
   | TestDone| TestFail
   | TestInterrupt| TestInvalid
   | TestMode| TestPortSelect
   | TestRun| TestWrapperControl
   | Unused| UnusedDuringTest
   | User USER_DEFINED >

exec_enum =
   < Characterization| Diagnostic
   | Production| User USER_DEFINED >

pattern_or_burst_enum =
   < AtSpeed| ChainContinuity
   | CompatibilityInformation| Endurance
   | EstablishMode| IDDQ
   | LogicBIST| MemoryBIST
   | Parametric| Retention
   | Scan| TerminateMode
   | User USER_DEFINED >

procedure_or_macro_enum =
   < Capture| Control
   | DoTest| DoTestOverlap
   | Instruction| MemoryPrecharge
   | MemoryRead| MemoryReadModifyWrite
   | MemoryRefresh| MemoryWrite
   | ModeControl| Observe
   | Operate| ShiftIn
   | ShiftOut| Transfer
   | Update| User USER_DEFINED >

identifier_event_enum=
   < Capture| Control
   | ControlObserve| DataFromCurrentActivity
   | DataFromPriorActivity| DataFromCurrentAndPriorActivity
   | Hold| Measure
   | Observe| TestPatternUnit
   | Reference| Transfer
   | Update| User USER_DEFINED >

relation_enum =
   < Common| Corresponding
   | Differential| Equivalent
```

```
    | Independant| InOutSet
    | MuxSet| OneCold
    | OneHot| Port
    | ZeroOneHot| ZeroOneCold
    | User USER_DEFINED >

ScanDataType_enum =
    < AddressGenerator| Boundary
    | Bypass| Counter
    | DataGenerator| Identification
    | Instruction| Internal
    | ResponseCompactor| User USER_DEFINED >

Environment (ENV_NAME) {
    ( CTL (CTL_NAME) {
        ( CoreInternal { } )
        ( DomainReferences ( CORE_INSTANCE_NAME)* {
            ( ScanCellGroups ( CELL_GROUP_NAME)+ ; )
            ( MacroDefs ( MACRO_DEF_NAME )+ ; )
            ( Procedures ( PROCEDURE_NAME )+ ; )
            ( SignalGroups ( SIGNALGROUPS_NAME )+ ; )
            ( ScanStructures ( SCAN_STRUCT_NAME )+ ; )
            ( Timing ( TIMING_NAME )+ ; )
            ( UsePrefix DomainNames; )
        } )* // end DomainReferences
        ( External { } )
        ( Family (NAME)+ ; )
        ( Focus ( Top) (CoreInstance (CORE_INSTANCE_NAME)+ ) {
            ( PatternTypes (PATTERN_OR_BURST_ENUM)+ ; )
            ( PatternBurst (PATTERN_BURST_NAME)+ ;)
            ( CTL core_instance_name CTL_NAME; )*
            ( TestMode (test_mode_enum)+ ; )
            ( Usage
                (<Production | Characterization | Diagnostic | User
USER_DEFINED>)+;)
        } )* // end Focus
        ( Inherit CTL_NAME; )
        ( Internal { } )
        ( PatternInformation { } )
        ( Relation { } )
        ( ScanInternal { } )
        ( TestMode (test_mode_enum)+ ; )
        ( TestMode (test_mode_enum)+ {
            AlternateTestMode (ctl_name)+ ;
        }) // end of TestMode
        ( TestModeForWrapper WRAPPER_TEST_MODE
TEST_MODE_CODE;)
        ( Vendor (NAME)+ ; )
        ( Wrapper < IEEE1500 | None | User USER_DEFINED > ; )
    } )* // end CTL
} // end Environment

Environment { CTL (CTL_NAME) {
    Internal {
        ( sigref_expr {
            ( DataType ( data_type_enum )+ ; )
            ( DataType ( data_type_enum )+ {
                ( ActiveState
                    < ForceDown
                    | ForceUp
                    | ForceOff
                    | ForceValid
```

```
    | ExpectOff
    | ExpectLow
    | ExpectHigh
    | ExpectValid > (Weak); )
  ( DataRate {
    ( NumberEvents INTEGER; )
    ( NumberVectors INTEGER; )
    ( NumberPatterns INTEGER; )
  } )
  ( AssumedInitialState
    < ForceOff
    | ForceLow
    | ForceHigh
    | ForceValid
    | ExpectOff
    | ExpectLow
    | ExpectHigh
    | ExpectValid > (Weak) ; )
  ( ScanDataType ( ScanDataType_enum )+ ; )
  ( ValueRange INTEGER INTEGER ; )*
} )* // end DataType
```

```
( DisableState
  < ExpectOff
  | ExpectLow
  | ExpectHigh
  | ExpectValid
  > ( Weak ) ; )
```

```
( DriveRequirements {
  ( TimingNonCritical; )
  ( TimingCritical {
    ( Period < Min | Max > time_expr; )*
    ( Pulse < High | Low > < Min | Max > time_expr; )*
    ( Precision time_expr; )
    ( EarliestTime time_expr; )
    ( LatestTime time_expr; )
    ( Reference sigref_expr {
      ( SelfEdge < Leading | Trailing | LeadingTrailing >
(INTEGER); )
      ( ReferenceEdge < Leading | Trailing |
LeadingTrailing > (INTEGER); )
      ( Hold time_expr; )
      ( Setup time_expr; )
      ( Period real_expr;)
      ( Precision time_expr; )
    } )* // end Reference
  } ) // end TimingCritical
} ) // end DriveRequirements
```

```
( ElectricalProperty
  < Digital
  | Analog
  | Power
  | Ground
  > (ELECTRICAL_PROPERTY_ID) ; )
```

```
( InputProperty
   ( < Edge
   | GlitchFree
   | PullDown
   | PullUp
   | Relation relation_enum
   | ScanStable
   | ScanUnstable
   | SynchFF
   | SynchLatch
   | Transitions INTEGER
   | Window
   | User USER_DEFINED
   > )+ ; )

( IsConnected <In|Out> (<Direct | Transitive>) {
   ( CoreSignal (sigref_expr) ; )
   ( IsEnabledBy Logic logic_expr {
      ( LOGICSIGNAME {
         Type
          < Signal | StateElement Scan | StateElement
NonScan | CoreSignal >;
         Name <SIGNAME | CELLNAME | SYMBOLNAME>;
      } )+ // end logicsigname
   } )* // end IsEnabledBy
   ( IsEnabledBy < Macro | Procedure > NAME ; )*
   ( TestAccess (<
      Control
      | Observe
      | User USER_DEFINED >)+
      < Macro | Procedure > NAME; )*
   ( < LaunchClock | CaptureClock > SIGNAME {
      ( <LeadingEdge | LevelSensitive | TrailingEdge> ; )
      ( <Direct | Transitive>; )
      ( StateAfterClock <
         Connection
         | ExpectLow
         | ExpectHigh
         | ExpectUnknown
         | ExpectValid
         | Hold
         | Invert
         | ShiftState
         | User USER_DEFINED > ; )
   })* // end LaunchClock
   ( Signal (sigref_expr) ; )
   ( StateElement <Scan | NonScan> (cellref_expr) ; )
   ( Transform {
      ( WaveformTable (WFT_NAME)+; )
      ( Invert ; )
      ( WFCMap FROM_WFC -> TO_WFC; )*
   })* // end Transform
} )* // end IsConnected

( IsDisabledBy <In|Out> Logic logic_expr {
   ( LOGICSIGNAME {
      Type
      < Signal | StateElement Scan | StateElement
NonScan | CoreSignal >;
      Name <SIGNAME | CELLNAME | SYMBOLNAME> ;
   })+
} )* // end IsDisabledBy
```

```
( IsDisabledBy <In|Out> < Macro | Procedure > NAME ; )*

( < LaunchClock | CaptureClock > SIGNAME {
   ( <LeadingEdge | LevelSensitive | TrailingEdge>)+ ;
})* // end LaunchClock

( OutputProperty
   ( < Edge
   | PullDown
   | PullUp
   | Relation relation_enum
   | ScanStable
   | ScanUnstable
   | SynchFF
   | SynchLatch
   | ThreeState
   | TwoState0Z
   | TwoState1Z
   | Transition INTEGER
   | Window
   | User USER_DEFINED > )+ ;)


( StrobeRequirements {
   ( TimingNonCritical; )
   ( TimingCritical {
      ( Period < Min | Max > time_expr; )*
      ( Pulse < High | Low > < Min | Max > time_expr; )*
      ( Precision time_expr; )
      ( EarliestTime time_expr; )
      ( LatestTime time_expr; )
      ( Reference sigref_expr {
         ( SelfEdge < Leading | Trailing | LeadingTrailing >
INTEGER; )
         ( ReferenceEdge < Leading | Trailing |
LeadingTrailing > INTEGER; )
         ( Hold time_expr; )
         ( Setup time_expr; )
         ( Precision time_expr; )
      } )* // end Reference
   } ) // end TimingCritical
   ( DCLevelNonCritical; )
   ( DCLevelCritical {... tbd...} )
} ) // end StrobeRequirements

   ( ScanStyle < MultiplexedData | MultiplexedClock >
(LevelSensitive) ; )

   ( Wrapper <IEEE1500 | None | User USER_DEFINED >
      ( < CellID cell_enum | PinID USER_DEFINED_PIN_ID >) ; )
   })+ // end sigref_expr
  } // end Internal
}} // end Environment, CTL

Environment { CTL (CTL_NAME) {
   ScanInternal {
      ( cellref_expr {
         ( DataType ( data_type_enum )+ ; )
         ( DataType ( data_type_enum )+ {
            ( ActiveState
               < ForceDown
               | ForceUp
```

```
            | ForceOff
            | ForceValid
            | ExpectLow
            | ExpectHigh
            | ExpectOff
            | ExpectValid
            > ; )
    } )* // end DataType

    ( IsConnected < In|Out > {
       ( CoreSignal sigref_expr ; )
       ( StateElement <Scan | NonScan> cellref_expr ; )

       ( IsEnabledBy Logic logic_expr {
          ( LOGICSIGNAME {
             Type
             < Signal | StateElement Scan | StateElement
NonScan | CoreSignal >;
             Name <SIGNAME | CELLNAME | SYMBOLNAME>;
          } )+ // end logicsigname
       } )* // end IsEnabledBy
       ( IsEnabledBy < Macro | Procedure > NAME ; )*

       ( < LaunchClock | CaptureClock > SIGNAME {
          ( <LeadingEdge | LevelSensitive | TrailingEdge> ; )
          ( <Direct | Transitive>; )
          ( StateAfterClock <
             Connection
             | ExpectLow
             | ExpectHigh
             | ExpectUnknown
             | ExpectValid
             | Hold
             | Invert
             | ShiftState
             | User USER_DEFINED > ; )
       })* // end LaunchClock

       ( TestAccess <
          (Control
          | Observe
          | User USER_DEFINED )+ >
          < Macro | Procedure > NAME; )*

       ( Transform {
          ( WaveformTable (WFT_NAME)+; )
          ( Invert ; )
          ( WFCMap FROM_WFC -> TO_WFC; )*
       } )* // end Transform
     } )* // end IsConnected
   } )+ // end cellref_expr
  } // end ScanInternal
}} // end Environment, CTL


Environment { CTL (CTL_NAME) {
   CoreInternal {
      ( sigref_expr { // format = coreinstance:signame
         ( DataType ( data_type_enum )+ ; )
         ( DataType ( data_type_enum )+ {
            ( ActiveState
               < ForceDown
```

```
          | ForceUp
          | ForceOff
          | ForceValid
          | ExpectLow
          | ExpectHigh
          | ExpectOff
          | ExpectValid
          >;)
    } )* // end DataType

    ( IsConnected < In|Out > {

      ( CoreSignal sigref_expr ; )

      ( IsEnabledBy Logic logic_expr {
        ( LOGICSIGNAME {
          Type
          < Signal | StateElement Scan | StateElement
NonScan | CoreSignal >;
          Name <SIGNAME | CELLNAME | SYMBOLNAME>;
        } )+ // end logicsigname
      } )* // end IsEnabledBy
      ( IsEnabledBy < Macro | Procedure > NAME ; )*

      ( TestAccess
        (Control
        | Observe
        | User USER_DEFINED )+
        < Macro | Procedure > NAME; )*

      ( Transform {
          ( WaveformTable (WFT_NAME)+; )
          ( Invert ; )
          ( WFCMap FROM_WFC -> TO_WFC; )*
        } )* // end Transform
      } )* // end IsConnected
    } )+ // end cellref_expr
  } // end CoreInternal
}} // end Environment, CTL
Environment { CTL (CTL_NAME) {
  Relation {
    ( Common sigref_expr ; )*
    ( Corresponding sigref_expr ; )*
    ( Differential sigref_expr (sigref_expr) ; )*
    ( Equivalent sigref_expr (sigref_expr) ; )*
    ( Independant sigref_expr (sigref_expr (Set) ); )*
    ( InOutSet signame signame (signame); )* // enable, output
name, input name
    ( MuxSet signame sigref_expr sigref_expr; )* // enable,
mux(0), mux(1)
    ( OneCold sigref_expr; ) *
    ( OneHot sigref_expr; ) *
    ( Port sigref_expr (integer) ; )* // port data, port select value
    ( ZeroOneCold sigref_expr; ) *
    ( ZeroOneHot sigref_expr; ) *
  } // end Relation
}} // end Environment, CTL

Environment { CTL (CTL_NAME) {
  External {
    ( sigref_expr {
      ( AllowSharedConnection {
```

```
        ( Core NAME (sigref_expr) ; )*
        ( DataType (data_type_enum)+ ; )
        ( Family (NAME)+ ; )
        ( OutputFunction <And | Or | Xor> ; )
        ( Self sigref_expr ; )
        ( Vendor (name)+ ; )
      } )* // end AllowSharedConnection

      ( ConnectTo {
        ( Core NAME sigref_expr; )
        ( DataType (data_type_enum)+ ; )
        ( Fanout INTEGER; )
        ( Instruction; )
        ( NoRebuffering;)
        ( Symbolic SYMBOLIC_NAME; )
        ( TAM; )
        ( Termination
        < TerminateHigh |TerminateLow |TerminateOff
|TerminateUnknown >;)
        ( Safe; )
        ( WBR; )
        ( WIP ; )
        ( Wrapper
          <IEEE1500 | None | User USER_DEFINED >
          (< CellID cell_enum | PinID USER_DEFINED_PIN_ID >) ;
)
      } )* // end ConnectTo
    } )+ // end sigref_expr
  } // end External
}} // end Environment, CTL
```

```
Environment { CTL (CTL_NAME) {
  PatternInformation {
    ( < Pattern | PatternBurst > pat_or_burst_name {
      Purpose (pattern_or_burst_enum)+ ;
      ( Power power_expr < Average | Maximum > ; )*
      ( Fault { } )* // see below for Fault block syntax
      ( ForeignPatterns {
        FileName FILE_NAME ;
        ( BlockName NAME; )
        ( BeginLine LINE_NUM; )
        ( EndLine LINE_NUM; )
        ( BeginLabel LABEL; )
        ( EndLabel LABEL; )
      } ) // end ForeignPatterns
      ( Identifiers { } )* // see below for Identifiers block syntax
    } )* // end Pattern | PatternBurst

    ( PatternExec EXEC_NAME {
      Purpose (EXEC_ENUM)+ ;
      ( PatternBurst (BURST_NAME)+ ; )
      ( Power power_expr < Average | Maximum > ; )*
      ( Fault { } )* // see below for Fault block syntax
    } )* // end PatternExec

    ( < Procedure | Macro > PROCEDURE_OR_MACRO_NAME {
      Purpose ( procedure_or_macro_enum )+ ;
      ( UseByPattern (pattern_or_burst_enum)+; )
      ( Identifiers { } )* // see below for Identifiers block syntax
    } )* // end Procedure | Macro
```

```
    ( UseStandardTemplate < EstablishMode | TerminateMode
> {
        ( SequenceNumber INTEGER; )
        ( VARIABLE_NAME (WFCLIST)+; )*
    } )* // end UseStandardTemplate

    ( WaveformTable (WFT)+ {
        (Purpose (< Shift | Capture | Data | User USER_DEFINED
>)+ ;)
        ( WaveformChar (WFC)+
            (< Launch | Capture | LaunchCapture | Data>)
            (< Level | Pulse | DoublePulse | Complex >)
            ( < Critical | NonCritical >)
            ( Measure); )*
    } )* // end WaveformTable
  } // end PatternInformation
}} // end Environment, CTL

Identifiers (PATTERN_OR_BURST_ENUM)* {
  ( Variable VAR_NAME (values)+; )*
  ( EventType identifier_event_enum {
    ( < Label | Xref > LABEL_ID {
      ( < Prefix | Complete > ; )
      ( <Begin | During | End> ; )
      ( SequenceNumber INTEGER; )
      ( EventValue (time_expr)+; )
    } )+ // end Label | Xref
  } )* // end EventType
} )* // end Identifiers
```

```
Fault {
  ( Type
  < Toggle
  | StuckAt| StuckOpen
  | Transition| Path
  | Bridge| PseudoStuckAt
  | User USER_DEFINED > ;)
  ( Boundary < Cell | Primitive > ; )
  ( FaultCount integer; )
  ( FaultsDetected integer ; )
  ( FaultCoverage real_num ; )
} // end Fault
```

Pre	Descr	Mult
E	exa	10^{18}
P	peta	10^{15}
T	tera	10^{12}
G	giga	10^{9}
M	mega	10^{6}
k	kilo	10^{3}
m	milli	10^{-3}
u	micro	10^{-6}
n	nano	10^{-9}
p	pico	10^{-12}
f	femto	10^{-15}
a	atto	10^{-18}

Unit	Description
A	Amperes
Cel	Degrees Celsius
F	Farads
H	Henrys (inductance)
Hz	Hertz
m	Meter
Ohm	Ohms
s	Seconds
W	Watts
V	Volts

	Drive
D	ForceDown
U	ForceUp
Z	ForceOff
P	ForcePrior

	Expect
R	ExpectLow
G	ExpectHigh
Q	ExpectOff
M	Marker

	Unresolved
N	ForceUnknown
A	LogicLow
B	LogicHigh
F	LogicZ
?	Unknown

	Compare
L	CompareLow
H	CompareHigh
X	CompareUnknown
x	
T	CompareOff
V	CompareValid
l	CompareLowWindow
h	CompareHighWindow
t	CompareOffWindow
v	CompareValidWindow
S	CompareSubstitute
s	CompareSubstituteWindow

Op	Definition
min ()	minimum value
max ()	maximum value
MergedScan ()	boolean function that returns true to select BreakPoint
()	parenthesis
,	comma - used as a separator in user functions
	SI units & prefixes
.	dot used as a domain reference operator:
	xxx.yyy means item yyy within object xxx
::	double colon used as a domain reference operator:
	xxx::yyy mean item yyy within object xxx
!	negation (boolean value)
~	bit-wise negation
@	timing event reference
.	dot operator used as string concatenation (for long strings):xxx.yyy yields string "xxxyyy"
/	divide
*	multiply
%	modulus
+	add
-	subtract
<	less than (boolean value)

<table>
<tr><th>Op</th><th>Definition</th></tr>
<tr><td>>
<=
>=</td><td>greater than (boolean value)
less or equal (boolean value)
greater or equal (boolean value)</td></tr>
<tr><td>&&</td><td>and (boolean value)</td></tr>
<tr><td>||</td><td>or (boolean value)</td></tr>
<tr><td>==</td><td>equal (boolean value)</td></tr>
<tr><td>!=</td><td>not equal (boolean value)</td></tr>
<tr><td>&</td><td>bit-wise and</td></tr>
<tr><td>|</td><td>bit-wise inclusive or</td></tr>
<tr><td>^</td><td>bit-wise exclusive or</td></tr>
<tr><td>^~, ~^</td><td>bit-wise equivalence</td></tr>
<tr><td>?:</td><td>conditional expression</td></tr>
<tr><td>=
..</td><td>assignment
range operator (ellipsis)</td></tr>
</table>

Index

Standard Test Interface Language

Q

S

T

U